INTO THE IMPOSSIBLE:
THINK LIKE A NOBEL PRIZE WINNER

INTO THE IMPOSSIBLE

THINK LIKE A NOBEL PRIZE WINNER

LESSONS FROM LAUREATES TO STOKE CURIOSITY, SPUR COLLABORATION,
AND IGNITE IMAGINATION IN YOUR LIFE AND CAREER

BRIAN KEATING

COPYRIGHT © 2021 BRIAN KEATING
All rights reserved. All illustrations © 2021 Ray Braun Graphic Design.

INTO THE IMPOSSIBLE: THINK LIKE A NOBEL PRIZE WINNER
Lessons from Laureates to Stoke Curiosity, Spur Collaboration, and Ignite Imagination in Your Life and Career

ISBN   978-1-5445-2349-1 *Hardcover*
           978-1-5445-2348-4 *Paperback*
           978-1-5445-2347-7 *Ebook*
           978-1-5445-2350-7 *Audiobook*

TO MY FAMILY; MY WHOLE UNIVERSE.

CONTENTS

FOREWORD 9

INTRODUCTION 15

1. ADAM RIESS: THE STAR GAZER 30
2. RAINER WEISS: THE TINKERER 46
3. SHELDON GLASHOW: THE NUCLEATOR 62

   INTERSTITIAL... 77

4. CARL WIEMAN: THE TEACHER'S TEACHER.................82
5. ROGER PENROSE: THE SINGULAR MIND 94
6. DUNCAN HALDANE: THE ALCHEMIST 106
7. FRANK WILCZEK: A BEAUTIFUL MIND 116
8. JOHN MATHER: THE COLLABORATOR 134
9. BARRY BARISH: THE AVUNCULAR AVATAR 150

   CONCLUSION... 165

   ACKNOWLEDGMENTS 171

   ABOUT THE AUTHOR 175

FOREWORD

BARRY BARISH: CURIOSITY KILLED THE CAT BUT NOT THE SCIENTIST

What do the nine scientists in Brian Keating's book have in common, besides having a Nobel Prize? Perhaps the most interesting common attribute is their insatiable curiosities. In different ways, curiosity is the common driving force the interviewees articulate in their quests to understand the physical world. Each of these very successful scientists has been strongly driven to understand the unknown and the unknowable. Their very different strengths, weaknesses, and approaches to pursuing the frontiers of science and their own lives are revealed through selective articulation from Brian's probing interviews, accompanied by Brian's own very interesting and candid reactions and interpretations.

While reading this short book, don't skip the very interesting and short "Interstitial: The Scientific Method." What inspired Brian to write it is unclear, but its importance cannot be overemphasized. We deal with "alternate truths" and "fake news" on a daily basis. Aristotle taught us how to use inductive and deductive reasoning to advance knowledge, and Galileo introduced the use of experiments as a research tool. Finally, Newton, in the *Principia*, wrote down his four rules of reasoning, which established the scientific method. Now, we rely on statistical arguments to establish confidence in our experimental conclusions, as well as consensus, as emphasized by Keating. These same principles need to be applied to establishing the truth for societal questions, like global warming or the effectiveness and risks of COVID vaccines.

Lastly, I conclude with a personal observation. Understanding science is hard enough; understanding scientists is even harder. As a leading scientist, Keating deserves a lot of credit for also tackling the latter.

BARRY BARISH, LINDE PROFESSOR OF PHYSICS, EMERITUS, AT CALTECH, FACULTY MEMBER AT UC RIVERSIDE, AND WINNER OF THE 2017 NOBEL PRIZE IN PHYSICS

JAMES ALTUCHER: SKIP THE LINE
BY CHOOSING YOURSELF

When I was thrown out of graduate school in 1991, I didn't think that on the exact same day thirty years later, I would be writing a foreword to a book about Nobel Prize winners.

I don't feel qualified to write this foreword. I have not spent my life teasing out the secrets of the frontier of human knowledge. I am having a bit of the imposter syndrome at the moment.

I have not even had one single career, like physicist or chemist. I have often pursued success in many careers. Careers ranging from entrepreneurship to standup comedy and from writer to podcaster.

On my podcast and in my books, I've interviewed more than a thousand people I consider to be smarter, more successful, and more talented than I am. This is the great thing about having a podcast. I can call up a brilliant scientist, a world chess champion, the most talented athletes, and successful writers, and I get to ask them whatever I want.

I'm like a vampire who gets to absorb the lives of my guests. After I speak to, dare I say, a Brian Keating, I feel momentarily more brilliant. Like I could look out into space and pierce through its veil of secrets. But then that moment goes away, and astronomy turns back into astrology.

This is my job. I get to talk to people I view as the most successful I know—world champions in their fields, whether writing, chess, physics, or medicine—and translate the "what got you here" just enough so my readers can benefit and perhaps propel their own careers forward with the knowledge I obtained for them.

Wait, scratch that. I have to admit, I don't really care that much about my listeners. I am selfish. I interview these guests for my own purposes. So that I can be smarter, better, faster, healthier. Maybe I can scrape their brains like a match and set my own tiny fire on the planet.

I want to do what they do. Maybe I can be a physicist. Or a Grammy-winning songwriter. Or sell 100 million copies of my books (thank you, Judy Blume).

Brian puts up with me. On at least four occasions, I tried to pitch him different ideas of a grand unifying theory of physics. He laughs, and we continue the conversation. A conversation that ranges from topics such as the beginnings of the universe to the questions of "why are we here," "is there a god," and "is there anything that transcends the limited understanding we have of what meaning there might be to all of this?" This!

Physicists are our philosophers. They look away from their telescopes and try to put order into the absurd. The physicist is the ultimate absurdist. Trying to carve out bits of meaning in a universe, which at first glance (at most glances?) seems utterly devoid of meaning. And the more they view the world as absurd, the more likely they are to be propelled to the top of their fields. Those who go left when everyone else goes right will find themselves in new territory.

Brian is a much better vampire than I am.

He has translated the habits of geniuses into a language that I understand, that anyone can understand. What is it like to be the first in the world with a new thought? And then turn that thought into something real?

Brian Keating famously has not won the Nobel Prize. I hope he wins it one day. But it doesn't matter. The Nobel Prize is an outcome that means many things.

It's a cliché to say, "It's about the journey, not the goal." But I didn't realize how effective that cliche is until I read the unusual ways Brian's interviewees tripled down, quadrupled down on that process. Every moment of their lives, every question, even every setback (and there are many, although nobody was thrown out of graduate school, fortunately) moves these icons forward in their process.

To where? Where are they going? It doesn't matter. They don't seem to care. All they know is that they are going. They are doing.

In many conversations with Brian, I've learned that science is not about knowing the answers; it's about asking the right questions. Which sounds like a cliche. Maybe it's not that they ask the right questions but that they ask questions nobody ever asked before. It's okay if those questions are the "wrong" questions. They just keep asking.

Brian asked the right questions to these Nobel Prize winners, got the right answers, and then translated it for people like me in this book.

Will I become smarter as a result of reading this book? I can say with full ego and equipped with a healthy dose of Dunning-Kruger bias that the answer is yes. Yes I will.

JAMES ALTUCHER, HOST OF "THE JAMES ALTUCHER SHOW" AND AUTHOR OF THE NATIONAL BESTSELLER *CHOOSE YOURSELF*

INTRODUCTION

The only way of discovering the limits of the possible is to venture a little way past them into the impossible.

—ARTHUR C. CLARKE

When 2017 Nobel Prize winner Barry Barish told me he had suffered from the imposter syndrome, the hair stood up on the back of my neck. I couldn't believe that one of the most influential figures in my life and career—as a scientist, as a father, and as a human—is mortal. He sometimes feels insecure, just like I do. Every time I'm teaching, in the back of my head, I am thinking, who am I to do this? I always struggled with math, and physics never came naturally to me. I got where I am because of my passion and curiosity, not my SAT scores. Society venerates the genius. Maybe that's you, but it's certainly not me.

I've always suffered from the imposter syndrome. Discovering that Barish did too, even *after* winning a Nobel Prize—the highest regard in our field and in society itself—

immensely comforted me. If he was insecure about how he compared to Einstein, I wanted to comfort him: Einstein was in his awe of Isaac Newton, saying Newton "... determined the course of Western thought, research, and practice like no one else before or since." And compared to whom did *Newton* feel inadequate? Jesus Christ almighty!

The truth is, the imposter syndrome is just a normal, even healthy, dose of inadequacy. As such, we can never overcome or defeat it, nor should we try to. But we can manage it through understanding and acceptance. Hearing about Barry's experience allowed me to do exactly that, and I hoped sharing that message would also help others manage better. This was the moment I decided to create this book.

This isn't a physics book. These pages are *not* for aspiring Nobel Prize winners, mathematicians, or any of my fellow geeks, dweebs, or nerds. In fact, I wrote it specifically for nonscientists—for those who, because of the quotidian demands of everyday life, sometimes lose sight of the biggest-picture topics humans are capable of learning about and contributing to. Most of all, I hope by humanizing science, by showing the craft of science as performed by its master practitioners, you my reader will see common themes emerge that will boost your creativity, stoke your imagination, and most of all, help overcome barriers like the imposter syndrome, thereby unlocking your full potential for out-of-this-universe success.

Though I didn't write it for physicists, it's appropriate to consider why the subjects of this book—who are all physicists—are good role models. Physicists are mental Swiss Army knives, or a cerebral SEAL Team Six. We dwell in uncertainty. We exist to solve problems.

We are not the best mathematicians (just ask a real mathematician). We're not the best engineers. We also aren't the best writers, speakers, or communicators—but no single group can simultaneously do all of these disparate tasks so well as the physicists I've compiled here. That's what makes them worth listening to and learning from. I sure have.

The individuals in this book have balanced collaboration with competition. All scientists stand on the proverbial shoulders of giants of the past and present. Yet some of the most profound moments of inspiration do breathe magic into the equation of a single individual one unique time.

There is a skill to know when to listen and when to talk, for you can't do both at the same time. These scientists have navigated the challenging waters between focus and diversity, balancing intellectual breadth with depth, which are challenges we all face. Whether you're a scientist or a salesman, you must "niche down" to solve problems. (Imagine trying to sell every car model made!)

I wrote this book for everyone who struggles to balance

the mundane with the sublime—who is attending to the day-to-day hard work and labor of whatever craft they are in while also trying to achieve something greater in their profession or in life. I wanted to deconstruct the mental habits and tactics of some of society's best and brightest minds in order to share their wisdom with readers—and also to show readers that they're just like us. They struggle with compromise. They wrestle with perfection. And they aspire always to do something great. We can too.

By studying the habits and tactics of the world's brightest, you can recognize common themes that apply to your life—even if the subject matter itself is as far removed from your daily life as a black hole is from a quark. Honestly, even though I am a physicist, the work done by most of the subjects in this book is no more similar to my daily work than it is to yours, and yet I learned much from them about issues common between us. These pages include enduring life lessons applicable to anyone eager to acquire anew the true keys to success!

HOW IT ALL BEGAN

A theme pops up throughout these interviews regarding the connection between teaching and learning. In the Russian language, the word for "scientist" translates into "one who was taught." That is an awesome responsibility with many implications. If we were taught, we have an obligation to teach. But the paradox is this: To be a good teacher, you must also be a good student. You must study how people learn in order to teach effectively. And to learn, you must not only study but also teach. In that way, I also have a selfish motivation behind this book: I wanted to share everything I learned from these laureates in order to learn it even more durably. Mostly, however, I see this book as an extension of my duty as an educator. That's also how the podcast *Into the Impossible* began.

I've always had an insatiable curiosity about learning and education, combined with the recognition that life is short and I want to extract as much wisdom as I can while I can.

As a college professor, I think of teachers as shortcuts in this endeavor. Teachers act as a sort of hack to reduce the amount of time otherwise required to learn something on one's own, compressing and making the learning process as efficient as possible—but no more so. In other words, there is a value in wrestling with material that cannot be hacked away.

As part of my duty as an educator, I wanted to cultivate a collection of dream faculty comprised of minds I wish I had encountered in my life. The next best thing to having them as my actual teachers is to learn from their interviews in a way that distills their knowledge, philosophy, struggles, tactics, and habits.

I started doing just that at UC San Diego in 2018 and realized I was extremely privileged to have access to some of the greatest minds in human history, ranging from Pulitzer Prize winners and authors to CEOs, artists, and astronauts. As the codirector of the Arthur C. Clarke Center for Human Imagination, I had access to a wide variety of writers, thinkers, and inventors from all walks of life, courtesy of our guest-speaker series. The list of invited speakers is not at all limited to the sciences. The common denominator is conversations about human curiosity, imagination, and communication from a variety of vantage points.

I realized it would be a missed opportunity if only those people who attended our live events benefited from these world-class intellects. So we supplemented their visiting lectures with podcast interviews, during which we explored topics in more detail. I started referring to the podcast as the "university I wish I'd attended where you can wear your pajamas and don't incur student-loan debt."

The goal of the podcast is to interview the greatest minds for the greatest number of people. My very first guest was the esteemed physicist Freeman Dyson. I next interviewed science-fiction authors, such as Andy Weir and Kim Stanley Robinson; poets and artists, including Herbert Sigüenza and Rae Armantrout; astronauts, such as Jessica Meir and Nicole Stott; and many others. Along the way, I also started to collect a curated subset of interviews with Nobel Prize–winning physicists.

Then in February 2020, my friend Freeman Dyson died. Dyson was the protype of a truly overlooked Nobel laureate. His contributions to our understanding of the fundamentals of matter and energy cannot be overstated, yet he was bypassed for the Nobel Prize he surely deserved. I was honored to host him for his winter visits to enjoy La Jolla's sublime weather.

Freeman's passing lent an incredible sense of urgency to my pursuits, forcing me to acknowledge that most prize-winning physicists are getting on in years. I don't know how to say this any other way, but I started to feel sick to my stomach, thinking that I might miss an opportunity to talk to some of the most brilliant minds in history who, because of winning the Nobel Prize, have had an outsized influence on society and culture.

So in 2020, I started reaching out to them. Most said yes, although sadly, both of the living female Nobel laureate physicists declined to be interviewed. I'm incredibly disappointed not to have female voices in this book, but it's due to the reality of the situation and not for lack of trying.

A year later, I had this incredible collection of legacy interviews with some of the most celebrated minds on the planet. T.S. Eliot once said, "The Nobel is a ticket to one's own funeral. No one has ever done anything after he got it." No one proves that idea more wrong than the physicists in this book. It's a rarefied group of individuals to learn from—especially when the focus is on life lessons instead of their research. It would be a dereliction of my intellectual duty not to preserve and share them.

Still, if you read my first book, *Losing the Nobel Prize*—in which I criticize both the Nobel committee as well as the larger culture that has turned the Prize into a false idol, sometimes at the expense of more meaningful achievement—you may be surprised that I've now written an adjacent book with a much sunnier disposition. After *Losing the Nobel Prize* was published, some accused me of crying sour grapes. They said I was disingenuous, and that there is nothing I'd like better than to win my own Nobel Prize. I say that one can call for a reform of the system without suggesting we throw the system out; we could use the prestige and veneration to agitate for reform and to live up to what the Prize could be. On the other hand, I say, of course, the Prize already does represent something almost mystical to millions around the world, and those who have won it have much to teach and inspire. Why lose that opportunity? This is an important vehicle, as imperfect as it is.

Nevertheless, such responses to the book led me to engage

in some soul-searching about what the Prize really means to me. This certainly may have subconsciously drawn me back to the subject matter, fueling my desire to interview as many laureate physicists as possible. Regardless, I see no contradiction between my last book and this one because my criticisms have always been about the committee, never about the recipients. The process may be deeply flawed, but we can still use the tools and tactics of the winners to improve our lives. Further, as you'll see in these pages, none of these recipients were ever driven by ambitions to win the Prize, making them ideal role models. I wanted to know what it is about people who win the Nobel Prize that is worthy of emulation and imitation. And now I want to share that with you.

HOW TO APPROACH THIS BOOK

These chapters are not transcripts. From the lengthy interviews I conducted with each laureate, I pulled all of the bits exemplifying traits worthy of emulation. Then, after each exchange, I added context or shared how I have been affected by that quote or idea. I have also edited for clarity, since spoken communication doesn't always translate directly to the page.

All in all, I have done my best to maintain the authenticity of my exchanges with my guests. For example, you'll notice that my questions don't always relate to the takeaway. Conversations often go in unexpected directions. I could've rephrased the questions for this book so they more accurately represented the laureates' responses, but I didn't want to misrepresent context. Still, any mistakes accidentally introduced are definitely mine, not theirs.

Each chapter contains a small box briefly explaining the laureate's Prize-winning work—not because there will be a test at the end, but because it's interesting context, and further, I know a lot of my readers will want to learn a bit of the fascinating science in these pages, considering the folks from whom you'll be learning. Perhaps their work will ignite further curiosity in you. If that's not you, feel free to skip these boxes. If you're looking for more, I refer you to the laureates' Nobel lectures at nobelprize.org. There, you will find their knowledge. But here, you will find examples of their wisdom—distilled and compressed into concentrated, actionable form.

Each interview ends with a handful of lightning-round questions designed to investigate more deeply, to provide you with insight into what these laureates are like as human beings. Often these questions reoccur.

Further, you'll find several recurrent themes from interview to interview, including the power of curiosity, the importance of listening to your critics, and why it's paramount to pursue goals that are "useless." I thought about collecting like with like and pulling all of the themes into a separate chapter, but I feel the wisdom will resonate more powerfully if it appears within the context of each conversation. Like history, these chapters don't repeat, but they do rhyme. Feel free to read the chapters in whatever order you like. I have designed them to be read in order, but if you'd rather skip around, that will also work.

One last disclaimer: Although these interviews are with individual winners, no one did their work alone. Scientists work in teams. And the teams have only gotten bigger over the years, often crossing continents and spanning decades. That only three people per discovery are awarded Nobel Prizes, rather than entire teams, remains one of my biggest criticisms of the Nobel committee and process. As such, you'll see I've repeated the phrase "and team" ad nauseam—that repetition is intentional.

WHAT YOU'LL LEARN

You'll find no high-level physics here. There will be no equations or homework problems. In these pages are years of wisdom distilled into chunks of actionable intelligence, including examples of resilience, patience, and courage. You'll learn how to deconstruct the most vexing problems in your life, see common threads between widely separated aspects of your life or career and weave them together, and find meaning in the interactions in occasional struggles you have with collaborators along the way. You'll learn why it's essential not only to immerse yourself in the past of your craft but also to invest in the future of it by teaching upcoming generations of practitioners in your field.

You'll learn the virtue of patience, that science has a great deal in common with art, and the value in doing something for its own sake rather than to receive accolades and attention. And you'll be powerfully reminded to allow curiosity, beauty, and serendipity to bring joy into your life through the surprising cracks that open up each time we turn fresh eyes onto a new problem.

Why learn these skills from physicists specifically? First, they are problem solvers by design. They are also talented observers of physical reality, trained to minimize their biases. And they have done so by being generalists: by pulling tools from disparate fields, including mathematics, logic, art, and even mysticism. Finally, their ultimate goal is to make sense of the universe and our place in it, a goal all humans are eager to pursue. The scientific method is the most powerful tool to analyze the physical world around us. In that way, science belongs to all of us.

Perhaps most importantly, successful physicists—like the nine featured herein—all have excellent soft skills. They've had to learn how to communicate and lead, often through trial and error. When I ask my students the most important skills needed to be a physicist, they usually say mathematical ability or lab skills. Wrong and wrong. Communication skills and emotional intelligence are the two most important tools among the greatest minds in my field. Whether it's correlated or causative, the laureates in this book have the ability to recognize humanity and that physics is ultimately a science which can only be done by human beings.

But if you learn only one thing from this book, I want it to be that these geniuses are mere mortals. They suffer from the same foibles, challenges, and prejudices that afflict us all. Through these conversations, we can learn how better to deal with the afflictions ourselves.

Finally, even if you learn nothing at all, I believe you'll be inspired. I was directly affected by many of these laureates early in my career. And I was indirectly affected by others among them who had inspired my mentors. Inspiration is a chain—and my ultimate goal here is to lengthen and strengthen it.

THE CRUTCH OF GENIUS

There is a scene in *A Few Good Men* where Colonel Jessup barks at Lieutenant Kaffee, "You want me on that wall— you need me on that wall!" I've often felt that laypeople want to know that Nobel laureates exist more than they really want to know why they won the Prize! It's almost as if society sleeps better, collectively, knowing that such geniuses exist, perhaps if only to desist from doing the work themselves. It's a form of absolution and comfort, to some, to think, "Well, so-and-so may be a physics nerd, but they were lucky; they had some unfair advantage—genetic, birthright, status, or otherwise—that I do not have."

Or, as Nietzsche put it:

> Thus our vanity, our self-love, promotes the cult of the genius: for only if we think of him as being very remote from us, as a miraculum, does he not aggrieve us.... Genius too does nothing but learn first how to lay bricks then how to build, and continually seek for material and continually form itself around it. Every activity of man is amazingly complicated, not only that of the genius: but none is a "miracle."
>
> —FRIEDRICH NIETZSCHE, HUMAN, ALL TOO HUMAN: A BOOK FOR FREE SPIRITS

The laureates depicted herein—most, if not all, from humble backgrounds—built solid intellectual walls, showing that genius is often a triumph of hard work, not merely the caprice of fortune. To me, this is more comforting: what one craftsman can build, so can another. That is our task too. Brick by brick. Let us see how it is done.

Visit https://briankeating.com/think_book.php for free resources, to go deeper into the impossible, and to help me write the next chapter(s) in my story!

CHAPTER 1

ADAM RIESS: THE STAR GAZER

Adam Riess is a distinguished professor of physics at Johns Hopkins University and an astronomer at the Space Telescope Science Institute. In 2011, he shared the Nobel Prize in Physics with Brian Schmidt and Saul Perlmutter "for the discovery of the accelerating expansion of the universe through observations of distant supernovae." The work—

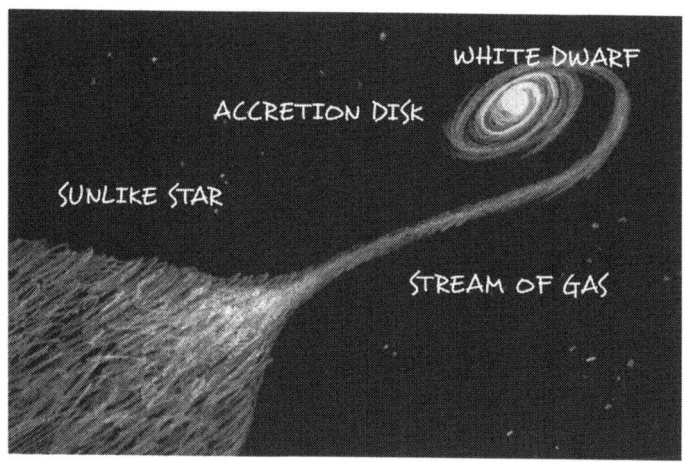

Here, a so-called 'Type Ia Supernova' is illustrated wherein one of the stars is a white dwarf accreting matter from another star, growing ever larger until it exceeds a critical mass (about 1.4 times the mass of our sun), creating an ultra-luminous detonation event visible across the entire universe.

done by a team—was recognized almost immediately (in Nobel years, at least), making Adam one of the youngest winners ever of the physics prize at age 41.

Though we are contemporaries, I consider Adam Riess a role model. His relentless pursuit of topics of great meaning is inspirational. In 2005, he and I competed in a worldwide competition to determine who is a worthy successor to the great physicist Charles Townsend, winner of the 1963 Nobel Prize in Physics. I won first prize for my concept for the BICEP experiment—spoiler alert: I did not succeed in replicating Charlie Townsend's renown—and third place went to Adam Riess. On the day he won the Nobel Prize, my brother Kevin said, "Brian, you won the battle, but he won the war"—something only a big brother would say.

I've learned from Adam that real winners never lose because they never give up. They're tenacious and insatiable. He's never rested on his laureate laurels. He is as hungry and passionate as ever. Winning the Prize at such a young age is kind of like winning the lottery—but instead of blowing it all in a year as many would, he has continued to work hard every day. His work also suggests that there's no such thing as a sophomore slump, which means we don't have to fear it. He inspires me. His dedication and curiosity and the good-natured aplomb with which he approaches his work are worthy not just of admiration but of emulation.

STAY CURIOUS
AMBITION ALONE WON'T SUSTAIN YOU

Keating: To what do you attribute your career success?

Riess: I've had wonderful mentors, and I've also been lucky to be at great institutions and work with fantastic facilities and colleagues—just learning the way science is done was so important. But for me, a very strong curiosity has been a driving force. I don't consider myself the smartest person in the field, but I make up for it by being fairly dogged in my pursuit of puzzles.

Keating: When did the astronomy bug first bite you?

Riess: It was a casual conversation I had with my dad when I was seven or eight. He was pointing out that stars are so far away that what we are seeing is really the way they were thousands of years ago, and that some of those stars may even be gone now. Just wrapping my head around this image of a star not being there but the light traveling anyway—it gets right into those juicy, fascinating aspects of physics.

Then, in terms of pursuing optical astronomy, that's where some of the most interesting questions happened to be at the time I was going to graduate schocl. Questions like, "How old is the universe?" Or, "What is the ultimate fate of the universe?" Who would think you could even begin to tackle questions like that!

Keating: In academia, we have what I call the Academic Hunger Games: you have to start off in a top-tier school, get good grades, and then impress your professors to get letters of recommendation, go to graduate school and get a good thesis project, publish papers, get first authorship, get a post-doctoral position, get a faculty job, get tenure, and eventually win the Nobel Prize. Do you ever feel like that game is broken?

Riess: I think that it's not a great scheme. I am lucky that I did not feel compelled by that scheme. Early on, I decided science was interesting to me and that I would pursue it as long as it was fun and engaging but that I wouldn't pursue it for checking certain boxes along some path. If people find themselves trying to check boxes—complete a list to get to a certain place—they're probably in the wrong field. Science is a lot of fun if you're driven by your own curiosity and passion. We do it because we want to know answers.

Adam's reference to childhood curiosity really resonates with me. If you are driven purely by ambition, that requires constant external validation and approval. Curiosity, on the other hand, is reinforced internally—it is self-validating. Your curiosity is unique. Only you can have it. And it is a good fuel to drive you to the stars. Following your curiosity is also a way of choosing yourself. It doesn't guarantee you a job, but it's much more sustainable and will provide you with much more resilience than constantly relying on external validation can.

A NOBEL IDEA
UNIVERSE ACCELERATION: WHAT IT IS AND WHY IT REVOLUTIONIZED PHYSICS

For millennia, scientists believed the universe was static. It seemed that, except for the planets, the sun, and the moon, nothing much moved in the heavens. All that changed in 1929 when Edwin Hubble showed 1921 Nobel Prize winner Albert Einstein and the rest of the world that all the galaxies in the universe were expanding apart from one another—envision raisins in an enormous raisin bread, each raisin representing a galaxy, all moving away from one another as the bread rises in the oven.

Astronomers expected that someday the expansion would stop, thanks to the irresistible force of gravity. In the 1990s, Riess and his team set out to measure how much the universe's rate of expansion was slowing *down*. That was the expectation. For example, if you throw a baseball, it will begin decelerating as soon as it leaves your arm: the gravitational force of all the mass in the Earth pulls on the baseball and slows it down.

Similarly, astronomers were well aware that the universe had vast amounts of matter—stars, planets, galaxies, and clusters of galaxies. They believed that mass had sufficient influence to cause the expansion of the universe to halt and start collapsing—that is, *de*celerating in its expansion. Further, we thought that eventually, the galaxies would slow down enough that they would turn around and come back together, perhaps crunching in on themselves in a universal collapse, a "Big Crunch."

But then Adam Riess and the corecipients of the 2011 Nobel Prize in Physics gathered data that led to the startling discovery that the expansion of the universe is in fact accelerating—faster and faster every day. This led them to interpret the existence of an additional component in the energy budget of the universe. Previously, we expected that matter, whether dark or luminous, is really all there was out there. Now, physicists believe about 70 percent of the energy in the universe is so-called dark energy—a notion originally theorized by Albert Einstein, and then discarded by him as his "biggest blunder."

The discovery of accelerating expansion is particularly remarkable because the team set out to measure the exact opposite phenomenon. I believe those are the purest discoveries, because they aren't as subject to the perils of confirmation bias. The team was awarded the Nobel only thirteen years after making the discovery, whereas most work waits decades to be reproduced enough to merit recognition. The immediacy of the acknowledgment led to rapid acceptance of the universe existing in an accelerating state, a view held as completely impossible just a year or two prior to the discovery.

BE AUDACIOUS
IF YOU'RE IRRITATING PEOPLE, YOU'RE PROBABLY ASKING THE RIGHT QUESTIONS.

Keating: Compared to other areas in physics, there is something about astronomy and cosmology that really excites the mind—and irritates it. Even Einstein is reputed to have said, when he first saw Lemaître's conjecture that there could have been a Big Bang, "Your calculations are correct, but your physics is atrocious." Why is astronomy so provocative?

Riess: Cosmology has a kind of audacity to address really big questions like, "How did this all start and where is it all going?" A lot of people feel that these are not questions for science, that they are questions for philosophy or religion. Cosmologists address these things. We don't say why there should be a universe or what one should do in the universe, but the kinematics, the motions, the distances—these are things we can measure.

Fred Hoyle, [the British astrophysicist, whom many believe should have shared in the 1983 Nobel Prize], got very upset at the idea of a Big Bang. He had a strong philosophical view about that. Cosmology draws us to the possibility of addressing those really big questions, which is where a lot of controversy can be as well.

I think it's so important to be audacious. It pairs well with confidence. You need to have confidence to be audacious, which is an essential item in the curiosity toolkit. If your work is irritating people, that probably means it's worth doing. It might not mean you're right, but it's a sign that you're asking the right questions. Just make sure to be judicious and ask questions in moderation.

STICK TO THE FACTS
BEWARE THAT YOU CAN BE MORE EASILY FOOLED THAN YOU THINK

Keating: You played perhaps the first and foundational

role in bringing the discovery of the "Hubble tension" to the attention of cosmologists. What is the Hubble tension, and why does it matter?

Riess: When we actually measure how fast the universe is expanding today, we get a number consistently and significantly higher than the prediction based on our understanding of the universe. It's like predicting how tall a child will be based on several height measurements, but then the child winds up two feet taller than you expected. Is this going to tell us about a new wrinkle in the universe? I don't know. It's very hard even to understand. Many of us continue to work on improving the precision of the measurements to parse those different possibilities.

[Tensions like this] give infinite rope for people to cherry-pick a piece of data and do a funky analysis. I'm a big believer in not censoring the data in that way. I think we learn the most by letting the data guide us through the universe, because our ideas have not been very successful ab initio [from first principles].

Being data-lead, whether in science or anywhere in life, is the best way to avoid confirmation bias. There is this notion that scientists are somehow immune from the very biases that make human beings human, which of course is not the case. I think often of a quote by Nobel Prize-winning physicist Richard Feynman: "The first principle is that you must not fool your-

self—and you are the easiest person to fool." Adam adheres to this guiding principle to have a healthy skepticism about all scientific conclusions, and especially your own. While living in the internet age with a lack of gatekeepers, it is more important now than ever before to be vigilant about avoiding confirmation bias.

LISTEN TO YOUR CRITICS
THEY'LL LEAD YOU TO YOUR GOAL

Keating: A statistical error—a random fluctuation, by definition—is unlikely to be reproduced the next time you or someone else does the same experiment. Therefore, it will be reduced with more data. But a systematic error that's intrinsic to your system or even the cosmos itself, such as emission from galactic dust, is very profound. How do we get rid of systematic errors?

Riess: You give talks and you listen to your colleagues. They generally ask good questions. Instead of dismissing them, you make a little note and sit around thinking of ways to answer your colleagues. It might seem at the time—especially when you're first starting out in research—that these are unfair questions. But you will do well to listen to those critics and try to come up with an experiment or test to answer them. If it succeeds, you go back to them and ask, "Is this convincing or not?" You continue the process until, as we like to say, it's true.

Carl Sagan said, "Extraordinary claims require extraordinary evidence." I've been fortunate to be around a few results that required extraordinary evidence, and it's hard. It's more than any one investigator, team, or group can do. It takes the community firing at things in different ways. And the community is pretty good at digging into these things. It's rare, in my experience, that something flat-out wrong survives that onslaught very long.

What people should understand is that cosmology is hard. The data are difficult to acquire and analyze. What leads to consensus in our field is when many different kinds of data that probe the universe in very different ways come to the same conclusion. This is part of the science process: we are able to reproduce reality, whereas a mistake or an error here or there is generally not reproduced in data set after data set. It's this reproducibility that leads us to think we are on the right track—and that defines what is special about science.

This part of our conversation resonated with me in several ways. First of all, Adam's words about statistical and systematic errors remind me of the serenity prayer: "Grant me the serenity to accept the things I cannot change, the courage to change the things I can, and the wisdom to know the difference." There are things in life we have no control over, like noise or statistical errors, but which can be reduced as we gain more data and experience. Over time, we improve but never quite get perfect.

On the other hand, there are things in life we have control over, such as systematic errors that are not only hard to remove but can be hard even to discover. But once you discover them, they can be removed. And that is like forgiveness—a feeling which borders on the divine!

This is why it's so important, as Adam says, to listen to your critics. Whatever you do for a living, you have to expect criticism. You shouldn't let the accolades go to your head, and you shouldn't let the criticism go to your heart. Expect, as a human being, that obstacles will appear on your path. As bestselling author Ryan Holiday pointed out, obstacles show you the path. They point to the direction of your goal. In this way, earnest criticism can be motivational.

INSIDE A NOBEL MIND
FINDING THE HUMAN IN THE GENIUS

Do you ever suffer from the imposter syndrome?

Riess: Early on, I realized that most of us in science only become quite knowledgeable about a very narrow niche. So I was able to recognize that it's really just that niche I'll know and not the broader picture. When the Nobel Prize happened—first of all, it was never really my goal or target—I quickly realized that it's not an IQ test or a ranking of great physicists. It's for people who, for the most part, were lucky that they were in the right place at the right time and contributed to a discovery. That could've been almost anybody in the field. I'm not Einstein and would never pretend to be. I'm satisfied having expertise in a tiny niche

What do you leave in your ethical will?

Riess: The best guiding light from science, physics, and everything I've experienced is to be curious. Curiosity is the antithesis, unfortunately, of what many people in the world suffer from today, which is a kind of extreme ideology, where one says, "I know everything already, and I'm going to beat other people over the head until they know it too." It never works out that way.

Scientists go in with curiosity, open their eyes, and learn about the way the world actually works. Often, you're surprised! In our Nobel work, we were extremely surprised about what the universe was doing. As long as people in the future can hold onto that curiosity to go out every day, question their assumptions, and look to see what is real and what is rational, they'll be well served.

Harkening to Sir Arthur C. Clarke's famous novel, 2001: A Space Odyssey, *there are monoliths meant to be encountered by human beings when they're ready to appreciate them. If you had to make a one billion-year time capsule, what would you put on it or in it?*

Riess: There's this very compelling idea about storing all the seeds of every kind of plant that's ever lived. Consider the complexity of the process that evolution went through to realize each of those! I would put a Noah's Ark of the plant world into my monolith.

Sir Arthur C. Clarke also said, "The only way to discover the limits of the possible is to venture beyond, into the impossible." What would you tell a twenty-year-old Adam Riess to venture into? What seemed impossible to you at the time, but then you went ahead and did it?

Riess: I knew I found science fascinating, but I didn't think there'd be a career, an actual way to do science. I loved learning but figured at some point I would have to grow up and get a job. I guess I would say to myself, "Keep following your passion. Don't give up until the door's been slammed a hundred times."

KEY TAKEAWAYS

- Adam and his storied career are testaments to the power of curiosity and audacity. Sure, it's possible to attain a degree of accomplishment by following existing playbooks, keeping your head down, and not asking too many questions. But true success is reserved for those curious enough to ask challenging questions and bold enough to attempt to answer them.

- Imagine how much more productive we would all be if we could truly understand that our critics actually help us achieve our goals. It can feel impossible to see it this way, which is why I find Adam's specific framing of this argument to be so profound and helpful.

- Adam turns the imposter syndrome completely on its head. If everyone is unique in their niche, then logically, no one can be an imposter. These arguments are important to remember whenever we doubt our work or question our place...which for many of us happens daily.

CHAPTER 2

RAINER WEISS: THE TINKERER

Rai Weiss is a professor emeritus at MIT, where he also earned his undergraduate degree and PhD. He, along with Barry Barish and Kip Thorne, won the 2017 Nobel Prize in Physics "for decisive contributions to the LIGO detector and the observation of gravitational waves," work that was completed with the rest of their team of more than a

The Laser Interferometer Gravitational-Wave Observatory (LIGO) project made the first detections of gravitational waves from massive compact objects colliding together in the distant universe. LIGO is so sensitive that it can measure distortions in spacetime far smaller than the diameter of a proton! In September 2015, LIGO detected the collision of two black holes, each about 30 times the mass of our sun, some 1.3 billion light years from earth.

thousand scientists, engineers, technicians, and managers. He has won numerous awards for his pioneering work on both the COBE (NASA's Cosmic Background Explorer) and LIGO (Laser Interferometer Gravitational-Wave Observatory) projects. He is a member of the National Academy of Sciences, the American Physical Society, and many other professional societies.

Weiss is a charming and consummate storyteller. His playful exuberance is incandescent. And I appreciate how forthright he is, as I think it is important to talk about the business of large projects and admit to the shortcomings rather than only discussing them with reverent worshipful praise. I admire that Weiss is relentlessly passionate even into his eighties. Just as a black hole shakes up space-time every few years, Weiss continues to pivot and switch direction himself. It's impossible not to think that this tendency has been responsible for his continued fertility of mind. He is a big-picture thinker that loves to keep things fresh by shaking himself out of complacency.

TEACH YOURSELF TO BE A PROBLEM-SOLVER
HINT: DO SO BY SOLVING PROBLEMS

Keating: You had developed a trade by the age of fourteen, working with record players or, as they used to be called, phonographs. And of course there is an interesting corollary between the microscopic vibrations of a tiny needle on vinyl and the reverberations of space-time.

Weiss: Many who have gone into experimental physics got their experience by being the fixer in the house. They worked in a garage or as an apprentice to a plumber or an electrician. It teaches you how to problem solve. Noise is the thing that kills you. Why did I get interested in electronics? It had to do with the end of the second World War. By 1944 and 1945, so much stuff was coming back as salvage. A friend who ran a junk store on Cortland Street in New York, near where the World Trade Center was, told me, "We got a nice thing, it just came back from the South Pacific. You'll have to get the varnish off, but there's a whole radar set there. You want the oscilloscope?"

In my life, the thing that was important was music. But I did not have the discipline to learn how to play an instrument. Then there was a lucky coincidence of three things: the junk on Cortland Street, loudspeakers from a movie theater in Brooklyn that had had a fire, and radio becoming commercial. I built several things with those speakers and started a business that was never intended. You could listen to the New York Philharmonic on the radio, and it sounded like you were right in the hall! I would invite other immigrant parents who were interested in classical music and they would say, "God, this is fantastic. Can you build us one?"

The phonograph record was the last piece of it, and that was an unsolved problem. How do you get rid of that record hiss? All the cures I came up with were worse than the disease.

Keating: But it got you acquainted.

Weiss: Yes. I went to college to try to solve that problem. Then, in the middle of [my] junior year at MIT, I found a lab which looked like they could use an electronics technician, and it turned out they could. I dropped out of school completely and became a card-carrying union member for a couple of years.

Weiss was born in Germany, and his family fled the Nazis. He taught himself multiple skills before heading to college, then dropping out, and eventually coming back to physics a bit later. His early life in the wake of the Great Depression during World War II fascinates me—seemingly he had an idyllic childhood as a self-taught tinkerer, despite the events reverberating around the planet. He has never lost that mischievous childlike curiosity and imagination.

This suggests to me that success does not depend on whether your path is straight or meanders but whether you're accumulating skills every step of the way. No matter your field—and this is even true outside of work, in relationships, in hobbies, in your avocations—problem solvers succeed, especially those who approach problems with imaginative solutions. I hate to be the curmudgeon that says, "Kids these days," but I do respect Rai's background and its relation to his remarkable career in physics. Anyone can engender a problem-solving mindset in themselves, at any age. Rai proves it's never too late.

FAIL IN ORDER TO SUCCEED
EVERY FAILURE TAKES YOU CLOSER TO YOUR GOALS

Keating: How do you know when to shut off an experiment? We can always keep them going if we shovel in more money, but how do you know when to shut them off?

Weiss: I have several examples because I have failed in a few places! My first failure was in making a better clock.

While I was an electronics technician for Jerrold Zacharias, he was working on atomic clocks. He said, "I want to do an experiment. You with your clock will be on top of [the mountain], I will be down in the valley, and we will send signals to each other and measure the Einstein red shift." What happened is that this fancy clock he had invented to do this didn't work. He decided he wanted to quit that experiment because he had more important things to do. But what did I have to do? I wanted to find out why the thing did not work. So I did. Generally, I [only] give up on experiments when I convince myself it won't work.

A great scientist learns from failures. And they are never really failures if you have learned something useful. We can all apply this to our daily lives, as bestselling author James Altucher points out in Skip the Line, by doing small low-cost experiments. Say you want to pivot your whole career and do something completely new. Beforehand, you should do a low-cost, low-risk experiment that potentially circumvents the need to sink more into a particular new endeavor. These experiments should take no longer than it takes to actually learn something useful (unless what you've learned is that there is more to learn).

A scientist learns something each time. Succeeding is not the goal of an experiment. Learning is. You can design lots of small, low-cost, low-risk ways to test ideas before making dramatic changes. With each step, you become more prepared to eventually make that change—and you build intuition along the way.

A NOBEL IDEA
GRAVITATIONAL-WAVE DETECTION: WHAT IT IS AND WHY IT REVOLUTIONIZED PHYSICS

Gravitational waves are reverberations in the fabric of space-time itself. Einstein predicted them in his general theory of relativity, which was a massive upgrade to Isaac Newton's laws of universal gravitation. Einstein's theory shifted the notion of space and time to a four-dimensional playground "monkey bar set," in which all events in the universe take place. This theory had surprising serendipitous predictions all on its own, one of which is that space-time can ripple, shudder, and shake when massive objects—say the sun or a giant black hole thirty times bigger—are in motion. They produce waves of gravity that travel at the speed of light. The more massive the object and the faster it accelerates through space-time, the more energy is produced in the form of gravitational waves.

For the first time in history, the LIGO experiment directly detected evidence for gravitational waves traveling from a collision of two supermassive black holes, each one of which weighed about thirty times the mass of our sun. The collision happened 1.2 billion years ago and reached us in September 2015. Similar to the way Galileo opened up a new way to think about physics, astronomy, and cosmology by pointing an optical telescope toward the heavens in 1609, so too will LIGO's observations revolutionize our understanding of physics, astronomy, and cosmology through this new window into the fabric of reality.

DON'T BE TOO DEFENSIVE
BETTER TO ADMIT YOU'RE WRONG AND ACCEPT HELP THAN TO STAND ALONE WHILE CLINGING TO FALSE BELIEFS

Keating: There were a lot of stops and starts on the way to detecting gravitational waves. Even the founding father of the field, Joe Weber, was not immune to some of the pitfalls all scientists should be aware of.

Weiss: I knew Weber pretty well. He was a very good scientist. The trouble is he did not have enough ability to be self-critical. That is the fundamental trap he fell into.

Weber invented a lot of things that we now use for [work in] gravitational waves. He was into the idea that you want to look for a strain. He had it different than we have it, but it is the same story, just another way of talking about it. Then he had the idea that there would be a [certain] character to the noise and he had to have coincidence experiments—and we do the same thing. The people who worked with him at that time actually suggested to him to look into how [others] might detect gravitational waves. The sad part is that when it came to defending what he had, he did not do it in a way that most scientists would. He just said, "You did not do it my way." He did not discuss how he measured this sensitivity or say, "Let's compare the notes. How did you do it? How did I do it?" There was never collaboration, and it was quite hostile.

In science and in all endeavors, the inability to be self-critical is disastrous. Weber was an incredibly gifted scientist, but he didn't have soft skills. He didn't try to build consensus—he didn't know how to work the room, scientifically speaking. It takes more than just calculating general relativistic equations and building an ultra-sensitive detector. You must know how to interact with, motivate, convince, and lead people. If you don't already know, you need to learn, treating it as an essential part of your craft, whether you're a salesman or a nuclear physicist.

Further, he couldn't build consensus for his experiment because his experiment was flawed! He was absolutely correct that gravitational waves were detectable, but his approach was wrong. If he had just collaborated with other scientists, they might have all gotten there sooner. He couldn't, though. He was defensive. He didn't accept that the scientific method is the ultimate decider of who is right or wrong. Weber wouldn't relinquish his initial results to his dying day. If you're standing alone, it might be because you're a maverick. Or it could be because you're wrong.

EMBRACE YOUR RIVALS
IF YOUR GOAL IS TRULY AMBITIOUS, YOU'LL NEED THE BEST MINDS AVAILABLE

Keating: To eventually successfully detect gravitational waves, you put together what Doris Kearns Goodwin calls a "team of rivals." Caltech and MIT are two institutions that share a historic rivalry. You guys were in competition for the most precious resources: minds and money! How did the project possibly succeed?

Weiss: It turned out to be extremely difficult. It comes down to one real problem: style of doing. We got into trouble, which had to do with MIT more than Caltech. I was ambitious because of the way MIT was looking at the whole field of gravitational-waves physics. When I started, MIT wanted no part of it, and I could not put graduate students on the prototype. Why? Because it was engineering. "There was not going to be any science coming out of it," I was told many times, so they were not going to risk a graduate student's life on a thing which did not look to them like it would be worth the trouble. There was no sympathy for black holes at MIT. I thought the only way I could stay in business at MIT and in this business was to immediately start pushing for a LIGO, a full-scale thing, because I could prove on paper that you could not do it with a small device.

[However,] the Weber thing had gone to the point where people were writing books about [gravitational waves] being a pathological science. When [Richard] Garwin saw that LIGO was being contemplated in the NSF, he sent a letter to the head of the physics division. It was not written in a nasty way, but it was written with a "if you are going to continue with [this, it] could embarrass the NSF..." tone. So they forced a summer study on us, and they gave us a very good review! The thing is, Garwin was colored by the arguments he had with Weber. We lived with that, and that is why we were being especially careful.

Barry [Barish] is a superb scientist and an organizer par excellence. Kip [Thorne]—this is something you probably don't know about Kip—has a very tactile sense about physics. He is a theorist, yes, and he does not make mistakes when he makes equations like I do. If he thinks it's anything, it is right. But he also has a wonderful feeling in his gut about the physics. Kip convinced us that there was something wrong with the way we thought about where the noise comes into the interferometer.

Keating: There is a lot of lore about LIGO, including a concern that the signal had been faked and produced by hacker. Did hackers get into the software?

Weiss: We could never prove that there was *no* hacker, but it looked easier to say nature did it.

Ultimately, they succeeded because the rival teams found a way to cooperate and keep moving forward. It helped that they deeply respected one another, even though they were technically competitors. Having the confidence to work with your rivals is an incredibly valuable soft skill. Arguably, a project this big and bold could only have succeeded if the two biggest competitors in the field joined forces to research together. But you can apply Rai's "teaming with rivals" approach in other situations too and ensure that the best solution is reached.

PASS THE BATON TO THE YOUTH
WHEN YOU LEAVE YOUR UNFINISHED WORK TO THE NEXT GENERATION, YOU ENSURE YOU'RE PART OF THE WINNING TEAM

Keating: We talk about black holes as if they are real; we sort of know they are. And you are one of the people responsible for the visceral feeling of reality that the idea of them induces. How do we know that we can take the leap from observing their exteriors to saying that they have singularities at their cores, beneath their so-called event horizons?

Weiss: That to me is the next most important thing that is going to happen in this field. If you see something or if you don't, that is not the end of the story. The next generation of people who go into gravitational-wave detection will push for something that can actually measure it, a Big Bang Observer.

This speaks to something I say throughout the book: science is cumulative. Everyone is standing on everyone else's shoulders. Weiss is certain that singularities will eventually be found, and possibly even soon. His enthusiasm about that is a great reminder that even if you don't personally achieve everything you want in your work, you can take solace knowing that you're contributing still, by building up the next generation of people who will reach the finish line for you. In that way, you still participate in the achievement. This is also literally what happened to him with COBE (the Cosmic Background Explorer, which launched in 1989 to observe three properties of the cosmic microwave background): it was his successors to the project, not him, who received a Nobel Prize for that work. (See Chapter 8 on John Mather.)

DO THINGS BECAUSE THEY ARE FUN
AND SAY "TO HELL WITH THE CRITICS"

Keating: You already had COBE, the discovery of something foundational that had a phenomenal impact on many people and resulted in Nobel Prizes for your protégés. And then you kept working on gravitational waves, which many people considered to be fringe and ultimately took forty years from conception to consensus. What gave you the courage? What can we learn from your experience of choosing to stick with a moonshot that could have been a big waste of time and money—especially after being in the spotlight as a big success?

Weiss: Well, I do not think the way you are talking. Where an experiment ends, whether it is successful or not, is not the thing in my mind. What is in my mind? "Is it interesting?" That is the first question: am I going to do something which is interesting to me or even anybody else? The other one, which is more pedestrian: "Is it fun?" I found both [COBE and LIGO] fun and important, and they started in my lab at about the same time.

Keating: Can an experiment get so big that it loses that interest and fun factor that appeal to you so much?

Weiss: There was still within LIGO so many things that use your imagination and made you feel that if each succeeded, you would have another interesting thing to work on. There were many puzzles, endless puzzles that came up.

I think that perspective is wonderful: a big project has micro projects, and they all add up to the success of the whole project, but each micro project is fun in and of itself. At any moment during that time, LIGO could have been canceled. People thought gravitational waves were not detectable. Weiss showed grit, resilience, and perseverance. And he makes it sound like it was easy to not give up because it was interesting—because it kept being fun. When your work builds on what other people have done and you learn from others' mistakes, it never loses its novelty. This is one of the reasons experimental physicists—unlike mathematicians, it's said—get better with age: when your empirical toolkit never runs out of space, each project builds on the previous one, and there is always something new to tinker with.

INSIDE A NOBEL MIND
FINDING THE HUMAN IN THE GENIUS

Do you suffer from the imposter syndrome?

Weiss: I suffered it right away [at the Nobel Prize ceremony]. I remember looking at the King [of Sweden]. My deepest worry was about when I got to the king. He wasn't old and tottering, but he looked to me very uncomfortable with this heavy thing he is holding. How can we make sure that neither of us drop the medal? That was my biggest worry as I was walking toward the guy. Then, walking away from him, I said to myself, "This is ridiculous! How do I fit in the same group as Heisenberg or Fermi or you name it? I mean, this is complete nonsense."

Alfred Nobel left all his money to the Prize, plus he endowed what we call in Hebrew an ethical will, a "zava'ah." It means a wisdom will, what you want to leave your heirs. What wisdom that you have accumulated in your eighty-eight years on Earth would you leave to future generations?

Weiss: It is the advice that I give everybody, but I give it to myself also: if it isn't fun, get out of it.

Arthur C. Clarke said, "The only way of discovering the limits of the possible is to venture a little ways past them into the impossible." I want to know what advice you would give a twenty-year-old Rai Weiss from an eighty-eight-year-old Rai Weiss. Advice to your former self that seemed impossible back when you were a youth but now seems eminently doable because of your courage, insight, or sheer drive?

Weiss: That one is easy. You do not make anything up that is of very much value until you have been around to think a little bit about it. Also, make sure that you keep looking at the fresh ideas that occur to you, because some of them might actually be interesting. Do not just say it is too hard. If you think there is something there, it is worth your time. Every five years, question if what you are doing is still interesting to you or if it has become a habit. If you are in field X, and five years later you are still doing field X, you haven't asked yourself enough questions. At least ask yourself, "Am I getting what I like out of this? Is it still fun? Is this still interesting?"

KEY TAKEAWAYS

- Make time to tinker and be playful—in so doing, you learn how to problem solve, but, just as important, you learn how to fail and how to learn from failure. Follow what's interesting and fun to you. If it fits those two criteria, that's a sign that there are problems you can solve.

- Remember that you and your critics are often striving for the same goal. Embrace their criticisms. Sometimes, make them your partners. Additionally, the next generation coming up behind you is also striving for the same goals. Consider them collaborators and successors rather than rivals.

CHAPTER 3

SHELDON GLASHOW: THE NUCLEATOR

In 1979, Sheldon Glashow, Abdus Salam, and Steven Weinberg were awarded the Nobel Prize in Physics "for their contributions to the theory of the unified weak and electromagnetic interaction between elementary particles...." Born in Manhattan in 1932 to Russian immigrant parents, he is now a professor emeritus of physics at Harvard

CONDUCTOR ELECTROMAGNETIC FORCE

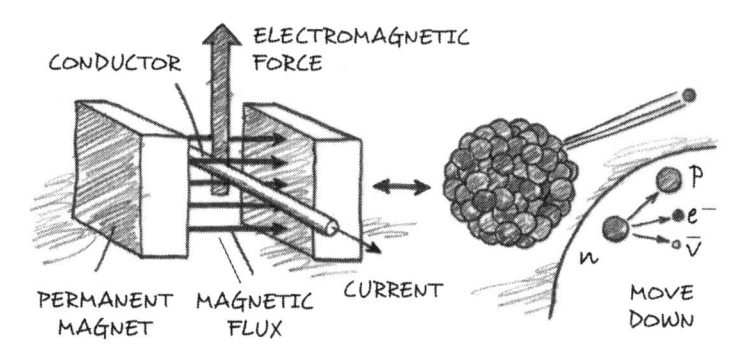

PERMANENT MAGNET   MAGNETIC FLUX   CURRENT   MOVE DOWN

The theory of the unified weak and electromagnetic interaction between elementary particles demonstrated that electricity, magnetism, and the weak nuclear force are part of a larger, more encompassing theory of fundamental forces and interactions.

University and professor emeritus of mathematics and physics at Boston University. Like a nexus, he's seemingly connected in one way or another to almost all of the amazing physicists of the twentieth century, and as such has both created nucleation sites for ideas to flourish and also catalyzed other interactions whose effects are still being felt today.

He is the classic figure of an American physicist and is allegedly the inspiration for the television character Sheldon Cooper on *The Big Bang Theory*. To be sure, Glashow is a bookish intellectual with a mischievous sense of humor. I admire that he is intellectually honest and rigorous but also playful and avuncular. He's cherished not least for his ability to describe the mysterious aspects of nature in a playful, delightful way. And all of his writings about science—including the wonderful book *Interactions*, published in 1988, which provided a pathway for the future of physics as he perceived it at the time—make room for the role of serendipity and luck in those discoveries that could never have been forecasted, no matter how many prizes a scientist might have won. He still has a sense of wonder in his late eighties.

PURSUE EXCELLENCE, NOT PRIZES
ANYWAY, THE FORMER MAY LEAD TO THE LATTER

Keating: Is rivalry anathema to theoretical physics, or could it be healthy if used properly?

Glashow: I have not had much experience with rivals in theoretical physics. It has always been, for me, a cooperative endeavor. For example, Steve Weinberg and I had our differences, but I would not call him a rival. He is just someone in physics with whom I had personal differences. No, it's been cooperative all the way.

Getting scooped is one of the greatest sources of anxiety for scientists. If someone publishes before you on the same topic, you can lose not only credit and the ability to dominate in the marketplace of ideas but also grants and career opportunities. Sheldon never had that fear. Even though it was clear to the rest of the physics community that he was in a race he easily could've lost, he displayed the opposite of competition throughout his career, always welcoming collaboration even when it risked robbing him of personal glory. This reminds me not to become so focused on attribution that I lose sight of the ultimate goal, which is to advance scientific knowledge.

Granted, we are only lauding this characteristic here because Sheldon did in fact win attribution and credit in the form of the highest prize in our field, the Nobel. However it is also true that everyone in this book worked just as hard after the Prize as they did before, suggesting that the Prize was not the goal. The pursuit of understanding nature has always been the goal. In the context of Sheldon, his personality is such that he doesn't take himself too seriously—and therefore competition is not a threat to him. This is something I try to do more of in my life.

MAKE TIME TO FANTASIZE
THOUGHT EXPERIMENTS ARE TOOLS

Keating: Carl Sagan wrote science fiction as well as being the foremost exponent of science nonfiction. Did science fiction influence you and your career in any way?

Glashow: Science fiction was an important part of my life. When I was twelve, thirteen, fourteen, I read [the magazine] *Astounding Science Fiction* religiously. For example, there was a column called "Brass Tacks," and it was there that I learned about the possibility of atomic bombs before the explosions took place in Japan. I can say that science fiction, to some extent, got me into science. I still appreciate it and will occasionally read it.

Pop culture has the power to stoke the imagination. Specifically, science fiction—when futuristic concepts and theories are contextualized via characters and relationships—frequently influences people to pursue actual science. The storytelling format allows us to explore ideas more fully, to dream big and let our imaginations run wild. Science fiction is a way to foretell the future, and such stories are serious. They lead to so-called thought experiments, exactly like those Einstein credited with his conceiving of the theory of relativity. He asked himself, "What would happen if I traveled alongside a light beam—what would I see?" Thought experiments don't need to be too strictly structured: the value of daydreaming is highly underrated.

A NOBEL IDEA

THE ELECTROWEAK THEORY: WHAT IT IS AND WHY IT REVOLUTIONIZED PHYSICS

Peter Parker and Spider-Man first appear to be different people but, upon closer inspection, are actually the same. Similarly, this work by Glashow, Salam, and Weinberg contributed to the understanding that two completely different forces of nature are actually two sides of the same coin. Namely, this became the unification of the electromagnetic force and the so-called weak nuclear force.

They showed that at very high energy or temperatures in the early universe, these two forces were actually one that later broke apart into two and today manifest themselves separately at relatively low temperatures. But during the early universe, you could've said that there was only one force. This is an example of what physicists call unification, taking things that seem disparate and showing through painstaking mathematical analysis that they're actually the same.

Together, they provided a pathway toward an ultimate theory of everything. This is one of the ultimate goals of physics, one Albert Einstein died never having realized himself: a theory that would result in an equation that would describe all of the interactions of nature as merely different manifestations of one underlying single force. In the 1860s, James Clerk Maxwell discovered that electricity and magnetism are actually different sides of one force, which we now call electromagnetism. Then, Glashow, Salam, and Weinberg showed that the weak nuclear force is yet another manifestation of that—meaning physicists have now unified what can otherwise be measured as three distinct forces. Next on the path would be to unite all of these forces with gravity as well. To date, that goal has proven elusive. But it has guided generations of physicists and continues to pull others who want to walk in the footsteps of scientists such as Glashow.

PLAY TO LEARN
TAKE A CUE FROM PRESCHOOL

Keating: I'm struck by how honest, how hilarious, and how magnanimous you are in your writing.

Glashow: I found doing science to be a lot of fun. I began doing real science as a graduate student. When I went off to Copenhagen [to spend two years at the Niels Bohr Institute], I discovered a plethora of post-doctoral students from many countries: China, Russia, Japan, Poland, Italy, Sweden. I wrote papers with them. I realized that their cooperation is the name of the game. And I was having a ball traveling. Fun is the name of the game in science, and it has always been that way. The "Glashow-Iliopoulos-Maiani" paper, which I am very proud of, emerged in part on the beaches of Mexico. Swimming around in the ocean, we came upon our idea. It was fun all the way along.

I've heard it cited that kids laugh three hundred times a day and adults only five. What happens between childhood and adulthood that makes us so serious? Perhaps, in our work, we can recapture some of that exuberance of childhood during what Jerry Seinfeld called "garbage time." In such unstructured moments, the mind ruminates. Sheldon unifies play and serious work. He drew inspiration from diverse intellects from around the world, and it was during unstructured playful time that their collaborative ideas emerged. Without those expansive, restful periods of play and fun, maybe Sheldon would not have been the physicist he was. We all need time to recharge and rejuvenate in between work. Otherwise, intuition has no room to bubble to the surface.

INSPIRE KIDS
IT'S A PATH TO HAPPINESS

Keating: Why do you call teaching "a secret weapon of the West"?

Glashow: It is the American tradition and the British tradition that researchers are teachers almost always. It is not necessarily true in countries like Germany, where there have been a great deal of researchers at institutes that do not teach. Germany has not suffered terribly from this procedure. I think teaching and research go together neatly, but they do not have to. I probably would have done even more research had I not had the responsibilities of teaching. But I probably would not have been as happy, because seeing how these kids learn, when they do, is very beautiful. I have had some wonderful students. That has been the most essential part of my life.

I love his candor here, saying that he probably would've done more research had he not been tasked to teach, but that he would not have been as happy. Teaching did not enhance Sheldon scientifically or help his career by leading to more breakthroughs and resulting prizes. But it brought him joy and human connection. More important, it inspired the next generation of students who will make the next ripples in the fabric of scientific space-time—which is perhaps a larger responsibility. For example, I only became aware of Sheldon's exceptional abilities as a teacher much later, from my friend Stephon Alexander, presently the President of the National Society of Black Physicists, who was deeply influenced by Sheldon at a young age. This influence in part led him to become a theoretical physicist, which in turn has influenced me deeply in my own work. In that way, Shelly's interaction on me was weak and long-range but, at the same time, dramatic and irreplaceable. Teaching and mentorship generate powerful waves. No one can say upon which distant shores they may break, only that the coastline of knowledge will be powerfully shaped by their having done so.

SEEK ELEGANCE
THINGS ARE OFTEN SIMPLER THAN THEY LOOK

Keating: Often, in physics and mathematics, elegant equations and beautiful symmetries lead us to truth. However, some believe we've become too intoxicated with beauty as a guide. Do you think there is a road too far where beauty can no longer lead us?

Glashow: I do not think so. I would follow the tradition of Einstein and others in advocating elegance and beauty. I anticipate there will be future discoveries of even more beautiful synthesis. We need to know so much.

Nature is elegant. When we're talking about beauty and physics, we are usually talking about simplicity and symmetry—the idea that truths are often uncomplicated. The poet Keats believed "truth is beauty and beauty is truth," but I'm not so sure. Sometimes beauty can be a good guide to help you winnow down possibilities when solving a problem and get to the truth more efficiently. Be aware of it as a tool, not intoxicated by it as a distraction. Look for elegance and simplicity and cultivate those in what you do. It might not always work out. Even when we can't solve problems or puzzles, we can appreciate and be fascinated by the many mysterious aspects of the world.

FOCUS FIRST ON UNDERSTANDING
NOTHING ELSE HAPPENS WITHOUT IT

Keating: What would you advise young physicists today to pursue?

Glashow: Personally, I prefer the useless sciences. Much of the research [you and I do] is in fact useless. Many of the wonderful discoveries that have been made will have no direct impact on our lives except the appreciation that we are understanding the world a little bit better each day. Our observations are a remarkable development of the past few decades. The fact that cosmology has become a precise science is very exciting. And there is so much more to learn about the universe.

The phrase "useless science" is Shelly's tongue-in-cheek way of describing so-called basic or fundamental research that won't necessarily lead to better cellphone coverage but might unlock secrets of the universe at its core. That's what makes physics and the fundamental sciences in general so important to our culture, not just to our technology. Sheldon is enchanted by the natural world. And that can be enough.

It's also true, of course, that most technological breakthroughs came out of traditional accomplishments in the fundamental sciences, from mechanical engineering to electronics and computers. The foundational research for these developments was all at one time "useless." Who's to say that Sheldon's work on the unification of the weak nuclear force won't one day be used to develop some technology? But even if it doesn't, the majesty and power of finding out how nature works is intrinsically worthy itself. This is one of the pinnacles of human evolution. Even if the work never becomes useful, understanding the majesty and power of how nature works is one of the most significant accomplishments humans can achieve. Paradoxically, the most important ideas in all of human civilization often appear completely useless at first.

INSIDE A NOBEL MIND
FINDING THE HUMAN IN THE GENIUS

What things outside of physics interest you?

Glashow: How could one not be interested in the new advances being made in virology, in developing vaccines? Some of the COVID vaccines are based on absolutely new technology, having to do with messenger RNA. They are not like the old-fashioned things at all! That stuff is very exciting.

What about life on other planets?

Glashow: I am absolutely convinced there is life on other planets. Maybe in this galaxy, but if not, we have billions of others to choose from. They will probably be intelligent life as well. Will we have the good fortune to encounter such life? That, of course, I do not know.

What do you think about the theory that we are actually living in some sort of advanced simulation and, in reality, have no agency?

Glashow: No, I do not think we are simulations. That is the realm of science fiction. On the other hand, I was amazed when computers first began to win at checkers. I said they would never win at chess. Then they demonstrated that they could win at chess, and I said they will never win at Go. Then they beat the best Go players consistently. So there is no question that the competence of computers is growing. There are those who feel that computers will become sentient and will become smarter than us. Certainly, if they do become sentient, they will be smarter than us. Will this ever happen? The answer in my mind is no.

On the Voyager satellite, there is a golden disc, a record that an alien phonograph can play. Carl Sagan convinced NASA to put this record on the satellite that contains sounds of planet Earth. The disc is meant to last for one billion years, and it is a sort of time capsule. If you had your own billion-year time capsule, what would you put in it?

Glashow: My hope is that human society will last a thousand years. I am not optimistic that it will. There are too many outstanding threats. Species tend to die off. There is no reason not to think that our species will not die off—and we can help that with nuclear weapons and with the greatest experiment ever performed: to dig up all of the fossil fuels on Earth and burn them and see what happens. What is happening does not suggest to me that there will be anyone around in one billion years, let alone ten thousand years. One thousand years: maybe.

KEY TAKEAWAYS

- Cooperation is useful despite its risks. Although competition is a fact of life, we do have some agency in the matter. We can choose to see rivals as collaborators. Doing so can be healthy for whatever the project, but even if not, it's healthy for ourselves.

- I'm struck by Sheldon's incredible, almost preternatural self-confidence. And I wonder if it's related in some way to his playfulness, if one fuels the other and vice versa in a recursive fashion. When we are confident in our work, we are more willing to be playful about it, certainly. Perhaps the act of play itself develops psychological skills allowing us to trust ourselves in a way that leads to more confidence. Even if not, both of these characteristics are worthy of emulation. If one leads to the other, even better.

INTERSTITIAL

THE SCIENTIFIC METHOD

If we are using these physicists as mentors, as guides to good practice, then I'd like to explain the process of that practice a bit. The building of consensus from experts in as many different fields as possible—who arrive at a similar conclusion—is the hallmark of good science. No conclusion is based on a single authority, as it might be in religion, or based on sheer popularity, as may be the case with pop artists or musicians. Neither, of course, is the scientific method determined by politicians. It is only determined by consensus via the adversarial process known as the scientific method.

There are two different ways to get this done. First, science can be accomplished via deduction, which means starting from a very general perspective and then working your way down to the more specific. Sometimes, this is called the top-down approach. In a deductive case, we might begin thinking of a theory that we want to explore in more detail, and from there burrow down to a more specific hypothesis that could be tested against either existing observations and data or proposed evidence. Evidence either does or does not confirm our hypothesis, which is the last step in the process of deductive reasoning.

Inductive reasoning works the other way around. Inductive reasoning begins with an observation or pieces of data that broaden out to generalization by observing patterns and formulating a hypothesis from the observed patterns. Once you have a provisional hypothesis, then you can create from that a theory. Deductive and inductive reasoning are equally valid; the process chosen only depends on which came first, the hypothesis or the observation.

Therefore, a theory should not be taken as something speculative or even, despite the common root word, hypothetical. In common vernacular, people say, "Well, that's just your theory." But in science, it's very powerful to call something a theory. It means it's been tested and confirmed, such as Einstein's theory of general relativity, which is one of our best tested of all theories.

In both the deductive and inductive approaches, the end goal is still consensus. The scientific method is a social consensus factory. No individual is smart enough to figure out scientific truths alone. We need a huge community to make sure no one runs off the rails.

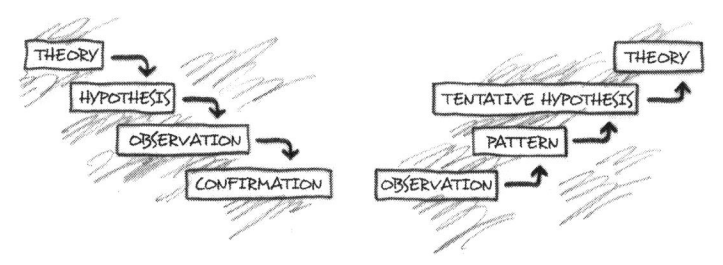

There is no, single accepted definition of the "scientific method", but two commonly adopted versions of it are shown here. The deductive version (left) follows a downward path from general to specific, starting with a theory and proceeding to an all-encompassing hypothesis. Example: the prediction of gravitational lensing and the bending of background starlight by massive objects between the observer and a source, such as the sun (predicted in 1915 by Albert Einstein as part of his Theory of General Relativity and observed by Arthur Edington in 1919 during a total Solar Eclipse). The inductive method (right) begins with an observation and proceeds "upwards" to a conclusive model or final explanatory theory. Example: the serendipitous observation of the Cosmic Microwave Background radiation by Penzias and Wilson, which ushered in the Big Bang theory as the dominant paradigm of cosmic genesis (see Mather's chapter). Both of these approaches are valid far beyond pure science! Leave an example of your favorite method on this book's webpage: https://briankeating.com/think_book.php

Despite the adversarial nature of the process (and, as I've explained, because of it), scientific debates actually come to a resolution quite frequently. In recent years, consensus has built around the Big Bang Theory versus its alternative, the Quasi-Steady State Theory, which is now perceived as likely to be wrong. To be clear, you can almost never prove something in physics to be right, but you can falsify it to the degree that it is almost certainly wrong. That's a very important distinction. The process of falsification itself must be present. If you can't prove something wrong—if there is literally no way to even try—then it doesn't matter what confirming evidence you have, because it's probably not going to be scientifically valid. It can't stand up to the scientific method. For example, you might be able to prove that the multiverse exists, but you can never falsify it. Someone could always say, "Well, there's another universe in the multiverse, but we won't ever see it because it's too far away." That means you can't prove it wrong, which also means you can't prove it right.

In the theory of human-caused global warming, on the other hand, consensus has been demonstrated because the conclusion has been reached from so many disparate angles. Evidence from people who study oceanography, paleoclimatology, geology, botany, and other disciplines give credence to the theory that humans are participating in the warming of the planet.

Consensus is how science gets done. It can only be reached through nearly endless testing and criticism. In science, you must not only trust your critics but accept the fact that you need them. Their participation as adversaries strengthens you and your work, making you anti-fragile. The more your theory is attacked and survives criticism, the stronger and more resilient it becomes. The scientific method teaches us to question our biases, seek criticism, and be less defensive in the name of a larger goal. It's the foundation of all of our work. And I find it quite beautiful.

CHAPTER 4

CARL WIEMAN: THE TEACHER'S TEACHER

Carl Wieman is a professor of physics at Stanford University, professor in the Stanford Graduate School of Education, and a DRC professor in the Stanford University School of Engineering. In 2001, he—along with Eric Allin Cornell, Wolfgang Ketterle, and their teams—was awarded the Nobel Prize in Physics "for the achievement of

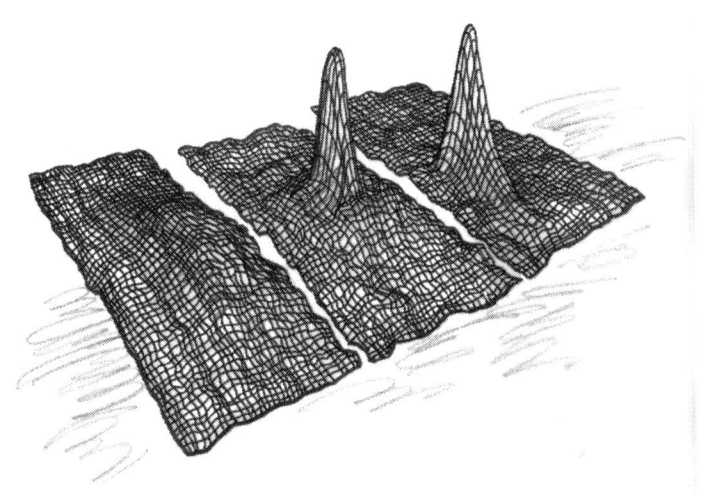

Here are shown three phases of the Bose-Einstein condensation (BEC) process. The leftmost panel illustrates matter above the critical temperature where the BEC is formed, the middle just at the critical point, and the far right when the new phase of matter—the BEC—is formed, showing sharp localization in space.

Bose-Einstein condensation in dilute gases of alkali atoms, and for early fundamental studies of the properties of the condensates." He was also a recipient of the 2020 Yidan Prize for education research.

His indefatigable work to revolutionize the way professors teach—and students learn—was the subject of our conversation. The first question I asked was, "If somebody says, 'I have good news for you and bad news for you,' which do you want to hear first?"

Without skipping a beat, he said he'd want the bad news because "it turns out negative feedback contributes much more to learning than positive feedback does." Wieman himself is dedicated to continued learning. This is only one of the qualities I find so inspiring in him. We are all educators and leaders, just in different ways, jobs, and positions. And it is impossible to be a good educator and leader without also being a good student. As Carl argues, one of the things we have to learn is how to teach. The field of teaching needs better best practices. We need to work smarter, not harder.

I admire the way Carl is now applying the same mental tenacity and clarity of purpose he used in the laboratory to his work in education. He realized that teaching is not dissimilar to the processes by which he conducted research—namely, that both are problem-solving exercises and involve certain hypotheses that need to be reexamined. As Carl points out, it's important to question assumptions and look at things in new ways. He argues that teachers have to be students and have to continue learning—and that they are not currently doing a good job of it. This is disruptive. The bad news: Carl equates current teaching styles to bloodletting. The good news: he doesn't think that teaching well will ultimately take much more time than we're currently spending.

ASK IF THERE'S A BETTER WAY
HOW THINGS HAVE ALWAYS BEEN DONE ISN'T NECESSARILY HOW THEY SHOULD BE DONE

Keating: Should we be teaching professors how to teach?

Wieman: It is very clear now that, yes, if you want to be a good teacher, there is a level of expertise you have to have. You have to know about the research into basic learning, and it connects with some basic cognitive psychology. And you have got to know how to implement those things properly in the classroom for the different range of students you have. In the thousands of years in which universities have been running, one did not have that research. It was kind of an individual art form.

The trouble is, we have not gone past that. A pretty good analogy is to think about medicine. We are at the place medicine was in during the mid to late 1800s. Somebody could come along and declare themselves to be a doctor; that was all it took. But then when you had scientific medicine coming along, you still had the kooks practicing their own idiosyncratic stuff at the same time real science was saying, "There's better ways to do this." So my soundbite on this is that the standard university professor is currently still practicing the pedagogical equivalent of bloodletting when there are antibiotics out there.

I am especially interested in this topic because I am working on my certified flight instructor training and have had to study a great deal of the theory of pedagogy for the first time, despite having been a professor at one of the top universities on the planet for more than seventeen years at the time of this writing!

One aspect of the pedagogical process that fascinates me is the role of human psychology in the process: what are the basic needs of a student? In the FAA's efforts to prevent people from dying on planes, they make sure their pilot instructors understand Maslow's hierarchy of needs as much as they understand how to land a plane if the engine suddenly quits.

Carl's take on the question of what bedrock principles teachers need to know makes perfect sense: of course teachers should know how to actually teach. Yet many in our field had not previously thought much about it. He's working to change traditional university teaching methods, such as the lecture format—a.k.a. the "sage on a stage" method held sacrosanct for over a thousand years—that he deems woefully and dangerously outdated.

Carl believes the act of teaching is yet another subject matter in which they need to have expertise. The standard professor never studies pedagogy and just parrots information they learned as students themselves. Carl points out how important it is to question the assumption that the way things have always been done is actually best practice. I think this is applicable to everyone, no matter your profession, because we all teach others: in training sessions, as mentors, and especially if you are a parent. Educating others is part of everyone's job. To be an effective teacher, you have to understand how people learn. The good news is that this really doesn't have to take that much time. More than anything, it requires an acknowledgment of the blind spot that kept us from seeing the answer to the riddle for so long.

A NOBEL IDEA
BOSE-EINSTEIN CONDENSATES: WHAT THEY ARE AND WHY THEY REVOLUTIONIZED PHYSICS

Carl Wieman and his fellow laureates created a new state of matter. In addition to the solid, liquid, and gas forms of matter known since antiquity, they created one called a Bose-Einstein condensate (BEC). It stemmed from predictions in the theoretical realm by Albert Einstein, based on a paper by Satyendra Nath Bose. The theory predicted that if you could cool a certain type of atom to very low temperatures, these atoms would behave kind of as a giant congregation of atoms with peculiar properties.

Almost eighty years after Bose and Einstein came up with this idea, Carl and his team were able to utilize technology to discover them. Forming these condensates requires incredibly low temperatures. In 1995, these temperatures were achieved for the first time, using a type of laser cooling that is able to reach very close to absolute zero, which made it possible for Carl and his team to observe the behavior of these condensates for the first time. Their work has had wide applicability in the study of atoms and molecules under extreme conditions.

IT'S QUALITY, NOT QUANTITY
EXPERTISE REQUIRES A VARIETY OF EXPERIENCES, NOT JUST AN AMOUNT OF TIME

Keating: What do you make of this so-called 10,000 Hour Rule—an interpretation of which was made famous by Malcolm Gladwell—that states, for example, that a master flight instructor needs to have accrued ten thousand hours of flight time?

Wieman: Yes, Gladwell was discussing the work of Anders Ericsson, who has done pioneering research into this. Certainly, ten thousand is an arbitrary number, but it is several thousand.

Keating: But with flight instruction, you really cannot encounter the different weather systems, terrain, air spaces (big cities versus small towns), and general conditions (day versus night) unless you do accumulate a lot of flight hours in different scenarios.

Wieman: That makes a very important point: Those hours have to be spent doing the right things. If you went always to the same place with the same weather, you could do ten thousand hours and you wouldn't become an expert pilot.

Just as a pilot couldn't learn anything new by flying to the same place with the same weather every time, in whatever you do, to grow, you must be testing and practicing in a variety of situations. You have to stretch. That includes not only research material required to be an expert in a subject matter but also an actual change in the way you teach. Growth involves pain and challenging yourself. Whether that's in the gym, the classroom, or in your job, you have to push yourself outside of your comfort zone. It's not enough to say, "I am smart, and therefore I can become a good _____." You have to spend a good amount of time learning the ins and outs of that vocation. Thankfully, according to Carl (and to Anders but maybe not to Malcolm), it won't take ten-thousand hours. Still, a home run doesn't count if you stop at third base.

REMEMBER THAT STRUGGLE *IS* PROGRESS
FORTUNE FAVORS THE PREPARED MIND

Keating: In my own education, there were different learning paths and tactics, but only some lead to big jumps and mental breakthroughs. Have you considered why that's sometimes the case for students?

Wieman: This is something cognitive psychologists have studied and I have wondered about myself. Their research looking at brain activity has shown that it actually is not *struggle, struggle, struggle, and then some great leap. It is develop, develop, develop, and then suddenly it becomes apparent.* Your brain processing has still been going on to reach the point you open the door. But the stuff before that door—all that thinking was actually prepping and wiring the brain in the right way. And then you completed the last link, as it were. It is important to keep that in mind. It's not, "Oh, I'm just waiting around." Your brain is actually turning; it has to be turning. People get frustrated thinking, "I am not really making progress." But you actually are. And when you have a great breakthrough, you should recognize, "All that great effort I thought was frustrating is not really frustrating." That is a problem when students are first learning something that is hard. They can very easily get discouraged because they have no idea how the process works.

I find this so comforting and motivating. Often, we have this notion that you have to literally wait for genius to strike. But in reality, you can open the door to genius by relentlessly preparing. It's like increasing the amount of surface area available for great ideas to stick. This is true whether you are technically a student or are simply continuing your daily education in your vocation or in one of your hobbies or side-hustles. Learning requires attention. If you think about it, that's really not too precious a thing to give—especially when it can have such huge payoffs. If a great mind like Carl's can continue dedicating time not only to teach but to learn *how to teach*—when he could be resting on his laurels, pun intended—that is an inspiration to me to be not only a better physicist but also a better teacher, in all the ways I am required to educate. My high school teacher Mrs. Tompkins always said that the word "educate" doesn't mean to pour into; it means to bring out of. Carl inspires me to bring the best out of myself and my students and children.

INSIDE A NOBEL MIND
FINDING THE HUMAN IN THE GENIUS

Can creativity be taught?

Wieman: I'm not going to talk about creativity in the arts, but I have thought a lot about creativity in science and have talked with people who study this. To be creative in science is basically where people look at some situation or question and simply find a way that's different than how everybody else has been looking at it. It's not bringing in something new. It's realizing things people already knew but did not really understand how to apply.

I would argue the standard educational approach we use in science is not *ineffective* in teaching creativity; rather, it is *anti-effective* in teaching creativity. In a normal course, students will learn things and be given tests they are graded on. The fundamental measure is always, "Are you able to produce the one answer the instructor wants to see?" That is completely the opposite of being able to think of ways to look at things or solve problems that nobody else has done before. You are penalized for creativity up through your entire formal schooling. You have to go through this big hurdle that says, "Get the answer all the faculty wants to see"—and that qualifies you to finally go out and do things where there isn't an answer. Stupid system, right?

To be creative in science seems irrelevant to the creative process in the arts. But actually, they're very similar, and the similarities extend to any other non-artistic field. You can achieve creative greatness in whatever your field is in the exact same way you would in the arts. Begin by reproducing the "masters." In so doing, you're effectively developing the muscle memory that an artist would get from reproducing the Mona Lisa *or Monet's* Water Lilies. *The muscle developed is not only physical dexterity but the mind as well. Some of your ten-thousand hours can be dedicated to that. You have a problem when all of your ten-thousand hours are dedicated to rote tasks such as memorization. Actually trying to reproduce the work of the giants who came before you instead of just memorizing it is a step toward mastery.*

KEY TAKEAWAYS

- Everyone has facets of their lives, no matter what they do for a living, that require them to be educators and leaders. Therefore, it's paramount to make time to learn exactly how people receive education and how people are effectively led. To improve as a teacher, you have to practice teaching. First, that requires you to have a student. But you must also study pedagogy—how to be a teacher—in order to communicate what you've learned. It's said that we only retain a small percent of what we read but much more of what we teach. That is the leverage, the secret return on investment that teaching gives back to the teacher!

- Then, as we seek to educate ourselves—whether in our own fields of study or in the ways of education itself—it's important to remember that the amount of time spent is not the best measure of what we have learned. We must seek out a variety of learning experiences and teachers in order to walk away with a more robust understanding. The famous ten thousand-hour metric is meaningless if you spend every hour on the same task or listen to the same lecture by the same teacher. Stretching builds muscles, mental and otherwise.

- Learning is hard. This is especially true if you're accustomed to picking up things pretty easily and more so when you don't have a good gauge of your progress. We can tend to equate the outcome with the process. But that's not how learning works. The mind is also growing and working during every step leading up to the breakthrough. Find a peer that you can bounce ideas off of, a sounding board. It's like getting batting practice—find a pitcher who's at your level, and together, you'll both grow. And make sure to reward yourself for each base you reach on the way to a home run, not just when you touch home base at the end.

CHAPTER 5

ROGER PENROSE: THE SINGULAR MIND

In 2020, Roger Penrose received the Nobel Prize in Physics "for the discovery that black hole formation is a robust prediction of the general theory of relativity," work he did along with his team in the 1960s. He is also a mathematician, philosopher, and the author of several books, including 1989's *The Emperor's New Mind*, an exploration

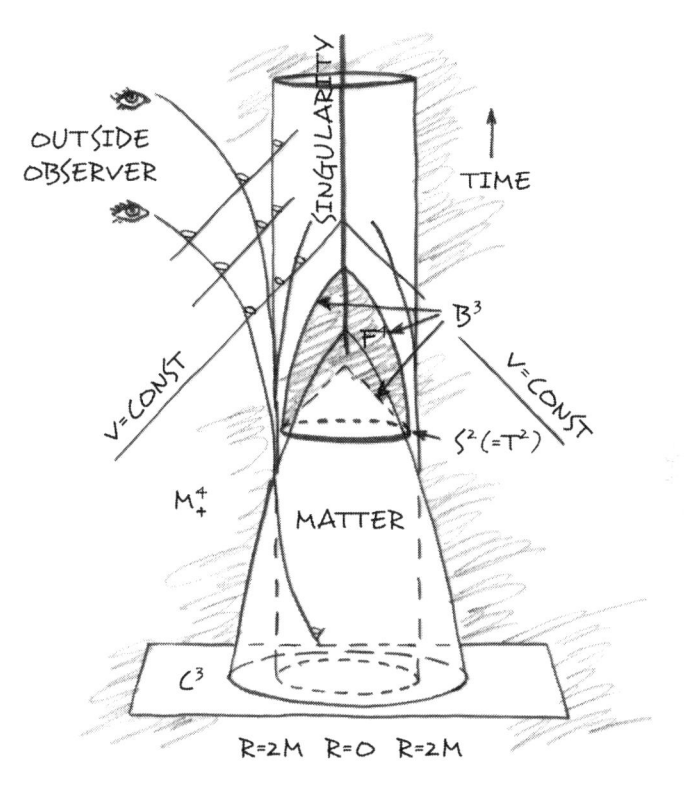

This diagram, adapted from Roger Penrose's paper "Gravitational Collapse and Space-Time Singularities" [Physical Review Letters Vol. 14 (1965), p. 57] illustrates the major phenomena observers could 'see' if they were unlucky enough to get too close to a Black Hole!

of consciousness and quantum mechanics, which not only had a profound influence on me in my youth but is also part of why I chose to write a popular science book myself. (That Roger Penrose blurbed my book was a thrilling capstone.) Today, he is Emeritus Rouse Ball Professor of Mathematics at the University of Oxford, among many other distinctions.

Not only is Roger Penrose relentlessly curious, but he's also fascinated by seemingly everything, from biology to black holes to consciousness. To the layperson, these might seem to all be similar esoteric theoretical activities, but to a physicist, they couldn't be more different. The fact that he can switch from a completely theoretical field, such as studies of the Big Bang, to a completely applicable one, like investigating the brain, is breathtaking to me. And he has had success in all of his fields.

Most people's work is either deep or broad. Roger's is both. It's the result of an incredible work ethic, a thorough imagination, and sheer longevity. He was eighty-nine years old when he won the Nobel Prize in 2020 and is still going strong. He doesn't give up, even when people consider his ideas on the outskirts of possibility. I deeply admire and try to emulate that kind of maverick self-determination to go where your curiosity takes you and your passions drive you, even if it's heterodox.

WORK WITH WHAT YOU'VE GOT
AN IMPERFECT TOOL IS BETTER THAN NO TOOL AT ALL

Keating: You've written before that innovations in engineering can be, like great works of art, discoveries of eternal truths as opposed to inventions. Do you feel this way about mathematics?

Penrose: If you're a mathematician, you strongly get the feeling that you are exploring a world out there, discovering things which are out there. If it works beautifully, there is something out there which is out of your control and it is much more like exploration. The more we learn about physics, [the more we see] it is governed by equations and geometrical ideas and we reduce them to mathematics. In that mathematics, we gain enormous precision in the way we describe and understand the way the physical world operates.

Penrose could not directly experience a black hole, the Big Bang, or a singularity. Instead of waiting for the perfect tool that would allow him to encounter these things, he used an imperfect approximation—mathematics—to work toward an understanding of the phenomena. And that was enough to get him the Nobel Prize. Abstract and therefore imperfect tools, such as art and math, among others, have merit in helping us discern the world. Don't let the lack of a perfect tool be the obstacle to reaching your goal. Persevere through other means—even if they don't get you all the way to the finish line, they may get you closer than you think. Perfect is the enemy of good.

DON'T FEAR THE DATA OVERLORD
FORTUNATELY, ALGORITHMS WILL NEVER BE ENOUGH

Keating: Is consciousness more than useful?

Penrose: The one thing I could say anything about is *understanding*. The ideas of computation lead me to believe that human understanding is not computable. It's not a computation. A lot of people argued with me because we have these wonderful computers and they do incredible things. I agree with that, but they run on algorithms. I think the argument is pretty clear that what we do when we understand a proof in mathematics is not following an algorithm. What is it in our abilities to think—to perceive conscious perception—that transcends computation? One argument people present is that the algorithms we use in our heads are so complicated that we will never be able to see. Sure, but the point is, how did that come about? By natural selection. What was naturally selected for was the basic principle of understanding. And that is not a computation.

People worry about the dangers of artificial intelligence. There's this panic that computers will take all of our jobs away—and then take over the world and destroy human life. But Roger argues in support of the animating spirit that differentiates humans from computers, providing solace by reminding us that there are things humans do which no computer ever could.

A NOBEL IDEA

BLACK-HOLE FORMATION: WHAT IT IS AND WHY IT REVOLUTIONIZED PHYSICS

Roger Penrose's discoveries of how black holes form validated a decades-old theory of Einstein's on general relativity. In the late 1960s, along with Stephen Hawking, Penrose proved that black holes collapse into singularities—regions in space and time where, basically, all hell breaks loose. In singularities, mathematics break down, as does the very nature of space-time itself. There's no way to understand them unless you employ the tools conceived by Penrose and his colleague and competitor, Hawking.

This work on black holes was highly mathematical. It's possibly the most purely mathematical of all the laureates' work featured in this book. Nevertheless, the tools conceived have allowed scientists to understand the phenomena of black holes—and perhaps even the origin of the universe—and thereby also test the theory of general relativity. Interestingly, Roger employed mathematical tools that were highly graphical—pictograms now known as "Penrose diagrams"—as devices to aid his studies of these most mercurial objects in the universe. These diagrams were highly innovative when first employed. Later, Penrose would again draw upon (no pun intended) his artistic skills when conceiving of "Penrose-tiles"; more on this below.

COLLABORATE WITH COMPETITORS
IT CAN ONLY MAKE YOU STRONGER

Keating: Did your work and Stephen Hawking's work intertwine in a way that felt like pure collaboration?

Penrose: He carried the arguments that I had [carried] originally a good deal further, and then we wrote a paper together. Certainly, there was a big influence in what he was doing, things he did afterward to do with the black holes. I thought he was doing the best work of anybody in general relativity. But we diverged views later on. He went off and started getting too influenced, in my view, by string theory and things like that. He tended to argue that black holes and white holes were the same, in some sense, which seemed to me to be absurd. He thought that space-time was somehow an observer-dependent concept, which was not the view I had. So we did diverge. Certainly, our disagreements were valuable to me. It did drive me in a certain direction. I had to think more deeply about things.

Stephen Hawking was a competitor of Penrose's in many ways, but they had a very productive competitive relationship, which set a strong example for the scientific community. They were at adversarial universities, Oxford and Cambridge, one of the deepest rivalries in cosmology and even in academia. Because of that, it's remarkable that they collaborated together. Some believe that if Stephen hadn't died in 2018, he would have shared a portion of the Nobel Prize with Roger. It's especially noteworthy that they continued to learn from each other even though their views, as Roger says, diverged. These competitors will forever be united by the work they did and the fact that they benefitted from mental sparring sessions, always conducted as proper British gentlemen should, of course.

Conversations between two collaborators can be more insightful, both to them and to the general public, when they are forced to defend opposing views. If you have respect for your debate partner and their opinion, you will endeavor to understand their ideas as well as your own, which will allow you to be self-critical, all of which will strengthen your understanding of your own viewpoint and your ability to defend it.

DIVERSIFY YOUR PURSUITS
CHANGING IT UP MIGHT LEAD TO BREAKTHROUGHS

Keating: What are the inspirations for your drawings?

Penrose: My old notebooks are full of these drawings—mostly they are where I got stuck, so I just drew up wild things.

Roger is known for his drawings, not least because they have enhanced science. He drew mesmerizing images—for example, non-periodic space-filling tiles, topologically mind-bending "Penrose staircases" and other M. C. Escher-like illusions thought impossible to manifest in nature. Nevertheless, some *variations of his images found a kind of real-world applicability. I think this speaks to the importance of intellectual diversity. To be clear, he does not credit his scientific innovations to his art, but many other physicists do make that connection for him. Some physicists regard his artistic accomplishments to be on par with ancient Greek scholars such as Pythagoras, whose musical explorations led to new physical and mathematical insights. Perhaps Roger's artistic flair played a role in the development of his intuition and discoveries in physics, or perhaps they didn't. Regardless, realizations in one medium or pursuit can certainly lead to breakthroughs in completely different genres.*

NEVER GIVE A FREE PASS
AND DON'T EXPECT ONE FOR YOURSELF

Keating: What excites you most at the moment?

Penrose: Roughly speaking, the crazy idea is that the very remote future connects to the Big Bang of the next aeon. And our Big Bang was the remote future of a previous aeon. Maybe it's just a guess, a speculation, and nobody would have proved me wrong. Then I had this idea that maybe collisions between supermassive black holes would produce signals that are strong enough that our curvature would influence it, which would affect the matter and you would see rings in the sky. That is what excites me most at the moment because you can see if the theory conforms to the observations or not. In the very remote future, most of the mass in a cluster of galaxies will get swallowed up by a black hole and eventually evaporated into photons. I used to think this is a very boring phase of the universe, when it has all evaporated away. Then I began to think, "Well, who is going to be bored by it? Only the photons!"

This theory, known as conformal cyclic cosmology, suggests that the universe goes through cycles of Big Bang and expansion, over and over again—and it is controversial. Roger acknowledges that and accepts that people aren't just going to trust everything he says now that he has a Nobel Prize. Instead of doing work that could never be disputed and would therefore allow him to preserve some amount of pride after having won the highest prize in his field, he is working hard to think of ways researchers could test his ideas—even though the result of those investigations could lead to increased skepticism of his theory. He never fell for the temptation to play it safe and live off of the reflected glory of his past accomplishments. This exemplifies both humility and courage, especially for someone of his age and accomplishment. Never trust anyone just because they've reached the highest position in their field. Likewise, never expect special treatment just because you're at the top of your field.

KEY TAKEAWAYS

- It's often said that mathematicians do their best work by the age of thirty. Sir Roger Penrose belies that claim. Considered by many to be the best mathematical physicist of the past century, he continues to work—and he is the oldest laureate in this book. Ingenuity, creativity, and sheer mathematical talent are necessary but not sufficient. Roger is indefatigable. Perseverance and tenacity may not get you everywhere, but without it, you won't get anywhere.

- Whenever Penrose hits a wall in his work, he finds a new way around it. Without any tools to measure or witness singularities, he turned to abstract theory at least as a way to continue thinking about the problems. He turns to art when he feels blocked—and then, amazingly, that art often leads to discoveries itself. When you're blocked, reframe the problem or approach it from a different angle.

- People often speak of Penrose and Hawking in the same breath because of their competition and collaboration. But there's another constructive comparison to be made between the two of them. Stephen was considered by many physicists to be a kind of a dilettante, hopping from one field to another (see the book *Hawking Hawking: The Selling of a Scientific Celebrity* by Charles Seife)—for example, from cosmology to string theory. Although Penrose also worked in different fields, he never abandoned one for the other. He stayed deeply connected to every field in which he studied. His work is and was both deep and broad. Having both perspectives no doubt enhanced his ability to innovate. We gain a more robust understanding of the world when we study it from a variety of perspectives, but only if each perspective is also thoroughly explored.

CHAPTER 6

DUNCAN HALDANE: THE ALCHEMIST

Duncan Haldane is the Sherman Fairchild University Professor of Physics at Princeton University. In 2016, he, along with David J. Thouless and J. Michael Kosterlitz, received the Nobel Prize in Physics "for theoretical discoveries of topological phase transitions and topological phases of matter."

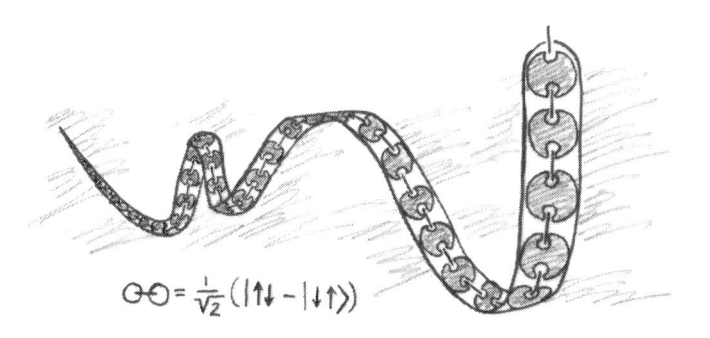

$\bigcirc\!\!-\!\!\bigcirc = \frac{1}{\sqrt{2}} (|\uparrow\downarrow\rangle - |\downarrow\uparrow\rangle)$

The theoretical discoveries of topological phase transitions and topological phases of matter explain how the spins of electrons (arrows pointing up and down) influence the magnetic properties of atomic chains in exotic materials. Haldane's work may someday lead to the development of novel electronic materials and even components for future devices that are unimaginable today.

There are many remarkable things about Duncan Haldane, not least his devilishly delightful sense of humor and his ability to playfully take on the most complicated matters. I also admire his intellectual curiosity, humility, and persistence. He is deeply driven by a desire to understand the possibilities of strange—even borderline bizarre—new forms of matter. Despite it taking many decades for his prize-winning work to be confirmed, he never gave up on it. At the same time, though, he never assumed he was right or allowed himself to be swayed by the notion that his brilliance would lead to a revolution in physics—as it has. The very day he learned he'd won the Nobel Prize, he reminded us of what really matters: he went right back to teaching and researching.

SEE THE BIGGER PICTURE
YOU MAY NEED TO GET SOME DISTANCE

Keating: Your work involved a chain reaction of hints and discoveries distributed among a variety of colleagues and universities.

Haldane: Everything has really come together through no one individual, but generally a recognition: something new showing up, and then it takes a bit of time to put it in context. It turned out that two things I contributed, I would not have said were connected in the obvious way. Only later, with the work of Xiao-Gang Wen, [did the connection become clear].

Duncan's work makes me reflect on our pursuits as grand tapestries. No matter your contribution, it's almost impossible to discern the overall pattern when you're up close to it. This is as true in theoretical condensed-matter physics—Duncan's line of work—as it is to those of us raising kids, coaching a sports team, or painting an actual portrait. Duncan realized that sometimes the pattern only emerges when you have enough distance, literally and figuratively. Only then can you take in and appreciate that you only contributed one thread to this glorious tapestry. And things that you would not have thought were obviously connected can turn out to be. Even if your work does not require you to collaborate directly with someone, you might be collaborating with somebody who died centuries earlier or be doing so unknowingly with someone who will come decades after you. The tapestry is never finished.

No matter your field, we all stand on the shoulders of giants and also provide shoulders for future generations. You can't know how your work might wind up mentoring someone years later. Therefore, the significance of your work cannot fully be appreciated at the time of creation. Duncan reminds us to be patient and put in the work anyway, allowing others to pick up the thread and continue unfurling it into the future. Rest assured that each thread is vital...even if you don't currently know what would happen to one if you pulled on it.

A NOBEL IDEA

PHASE TRANSITIONS AND TOPOLOGICAL MATTER: WHAT THEY ARE AND WHY THEY REVOLUTIONIZED PHYSICS

This work is some of the most theoretical of any physicist to win the Nobel Prize. In other words, it's difficult to explain. It was so difficult that the Nobel committee had to brandish props—bagels, donuts, and something called a "Swedish pretzel"—at the press conference when they announced the 2016 award. Frankly, almost no one understands quantum mechanics—not even some physicists, according to 1965 laureate Richard Feynman. But here's at least a quick rundown of why this work was so groundbreaking.

We'll start by way of analogy. Although you can visualize an ideal triangle in your head, you never encounter them in the world. What does a triangle "weigh"? It is a nonsensical answer. You can make a triangle, but doing so requires "stuff," and that stuff is technically not a set of three zero-dimensional points. True triangles don't exist. So too are there quantum mechanical entities that can be conceived of, but which, it was long thought, are impossible to produce in the lab. However, Duncan and his collaborators said otherwise. They predicted that these abstract entities *could* exist in nature if composed of highly engineered abstract matter. Later, that was shown to be true.

They demonstrated that the substance of nature isn't confined only to what the universe provides for us. Their work was one of the first predictions of a new type of matter previously unknown to exist and only able to be produced by human beings. It represents a grand partnership between humanity's genius and the collective ingenuity of vast networks of individuals—people working together to create something that was truly a novelty (whether or not it has any applicability to our daily life). Making new matter has been a goal of scientists for millennia, from ancient alchemists to chemists of the scientific revolution to those today. In that way, Haldane and his collaborators are a kind of alchemist. That is very cool.

SIMPLIFY THE PROBLEM
TAKE IT APART TO PUT IT TOGETHER

Keating: I remember hearing whispers that you guys

were going to win the Nobel Prize, that it was a foregone conclusion—

Haldane: I do not think it was a foregone conclusion. It was controversial. What made this Prize possible was the development by Charlie Kane, which pushed my work in a way I had not. And it took work by others, like Andrew Burnham Bacon and Xu Zhang. These things require three levels to come together. One, there are underlying abstract principles to be found, but they are difficult to understand. Second, the toy-model intermediary: you can actually do a calculation to see how all the bits fit together and maybe say something unexpected. Finally, the third piece is for someone to actually make a connection to physical materials, and then the thing takes off. Once real materials were found, then everyone was excited.

Duncan's starting point is the physicist's equivalent to a "killer app," and it's a hack you too can use. Physicists call it a gedankenexperiment, which is just a fancy way to say "thought experiment." Einstein was a master of such experiments, famously fantasizing what the world would look like if he could run at the speed of light. Now, if he actually waited till he could actually do such an experiment, we'd still be waiting for the theory of relativity! His key insight was to visualize the consequences if such an experiment could be done.

Now, if doing a low-cost, low-risk physical experiment in the real world is good, then doing a thought experiment—which is free—is even better. The starting point is to reduce the problem to its core essence. Doing so—creating what physicists call a toy model—may reveal ways to attack it that lead to fruitful ways of approaching the seemingly unsolvable problem.

In their case, they were reducing a problem to see how all of its constituent sub-problems could be deconstructed into simpler parts. By disassembling the problem, you might see how everything fits together, as Duncan did. And even better, once the answer is revealed in a simple case, it might inspire somebody else to look for more complicated cases—which might wind up helping you solve your initial problem. Solving the simplest version of your bigger problem can not only help you understand your bigger problem but might also encourage people to join in your endeavor.

DON'T BE TOO GOAL-ORIENTED
LEAVE ROOM FOR SURPRISE AND SERENDIPITY

Keating: How do you see this creative process leading to the beneficence for mankind envisioned by Alfred Nobel?

Haldane: The deepening of the understanding of nature, especially of quantum mechanics, is the seed corn for all kinds of future development of technology. I used to be skeptical that quantum computers were snake oil, but seeing how things improve when people start to get serious about looking at it, I think we will see some kind of advance of quantum information. I'm not sure what it will be or what form it will take. But Maxwell would not have predicted iPhones, right? So I think just getting a better understanding of the fundamental principles of how the world works is absolutely a benefit to mankind.

Sometimes the end result of a new endeavor is so far beyond the horizon that it can scarcely be envisioned. If you had asked James Clerk Maxwell (who lived in the mid-1800s) what would be the implications of his famous laws of electromagnetism, he would have had no idea. Yet almost everything in modern life is dependent on Maxwell's equations, from cell phones to the internet to lights in your home. People thought there would be no practical import of the technologies produced by quantum mechanics. Yet now all of modern computing is based on it. Physics teaches us the danger of forecasting the applicability or usefulness of research. Therefore, we should do basic fundamental research not for profit or a goal but for its own sake. And this is true for any endeavor. We must remember not to be so goal-oriented. Henry Ford said that if he had given people what they wanted, he would've tried to make a faster horse. Breakthroughs can't be predicted. That's part of their nature. Working too rigidly toward a goal can be counterproductive. There's value in work for its own sake because that's when we experience magical serendipitous breakthroughs that we could never have forecasted.

KEY TAKEAWAYS

- I always joke that serendipity is tough to plan on. It requires patience, showing up to do the work, day in and day out, without expectation of reward. Haldane labored for decades, not knowing whether or not the phenomenon he theorized would ever be discovered or if it even existed. His passionate curiosity kept him going. No matter how it turned out, it was an interesting problem to solve. That was enough. Try to get to a place where the work is its own reward. When you're working hard, you'll be surprised by how often good luck strikes.

- Remember that you are part of a bigger picture. People will build on your work. Maybe tomorrow or maybe in a few decades. If your work feels important to you, labor without needing to connect all the dots on your own or even necessarily in your lifetime. Pass the baton, just as it was passed to you.

- Try not to be overly concerned with the practical applications of your work. Even if practicality is your goal, focusing too much on it can have the opposite effect, stifling true innovation. Ford said if he had listened to his critics, he would have given them a faster horse. Think also about what you do from an artistic perspective, remembering that if people cared only about intrinsic value, the *Mona Lisa* would be worth the price of some paint and canvas.

CHAPTER 7

FRANK WILCZEK: A BEAUTIFUL MIND

Frank Wilczek is a physics professor at MIT, Arizona State University, and Stockholm University. He won the 2004 Nobel Prize in Physics, along with David Gross and H. David Politzer, "for the discovery of asymptotic freedom in the theory of the strong interaction." The work, which revolutionized quantum physics, was conducted thirty-

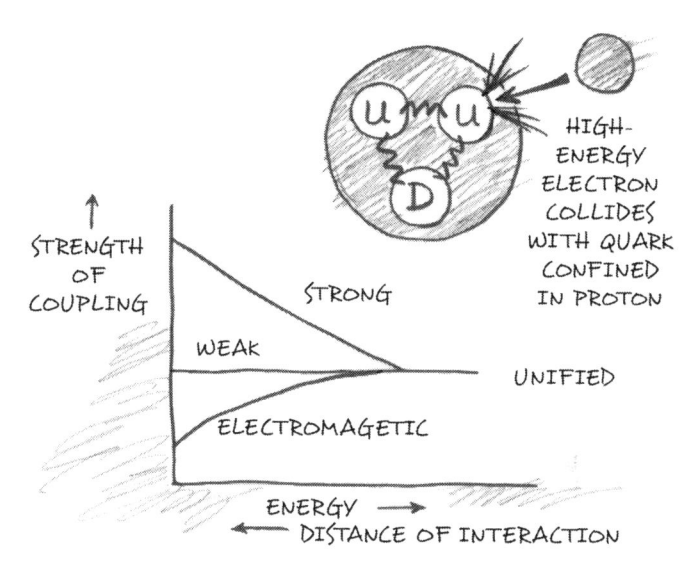

Here is illustrated an aspect of asymptotic freedom in the theory of the strong interaction: the inverse interplay between the energy and size scale at the subatomic (quark) level. So-called 'Grand Unified Theories' also predict that at high enough energies, the electroweak force (See Glashow chapter) will unify with the strong nuclear force, governed by quarks and gluons, the properties of which were revealed by Wilczek and collaborators.

one years prior, when Wilczek was a graduate student at Princeton. He has been awarded a MacArthur Fellowship and is a member of the National Academy of Sciences and the American Academy of Arts and Sciences. He has written multiple books, including *A Beautiful Question: Finding Nature's Deep Design* and *Fundamentals: Ten Keys to Reality*.

We tend to respect things in other people that we either can't see in ourselves or see in ourselves but don't live up to. This is probably why I so admire Frank Wilczek's endurance and forbearance. Those qualities are incredibly rare in a scientist and are ones I struggle with personally. He had this almost stoic patience, to wait thirty-one years for recognition, knowing all the while that he would likely win the Nobel Prize, but not receiving it. He exhibited incredible grit and resiliency. And he never lost his cheerfulness! Throughout those years, he committed to the process, dedicated himself to his craft, and kept showing up, understanding that the award is not the final arbiter of success or even satisfaction. I've learned more from him about patience and determination even than I have about the inner workings of protons and quarks.

REVERSE ENGINEER PROBLEMS
DON'T BE AFRAID OF EDUCATED GUESSES

Keating: The title to your book, *A Beautiful Question*, is a bit of a pun. You've said that it refers to an actual question: Does the world embody beautiful ideas? Can we use beauty not only in aesthetically pleasing endeavors such as art but also as a guide or tool in science?

Wilczek: Yes. The information we gather, for instance, on the backs of our retina, is not enough to reconstruct a three-dimensional world. We have to fill in a lot. So we rely on patterns. We rely on mathematical regularities—subconsciously, of course—and learn to as children. That's why it's not a matter of the world being designed to be beautiful. Rather, beauty is useful to us in understanding and harmonizing the world and being able to cope with it.

Symmetry is a useful guide; therefore, evolution has made us think of it as something desirable with which we want to interact. I turn the aesthetic version on its head.

When you want to understand the inner workings of atoms, it's much more difficult to do experiments. Instead, it turned out to be very successful to guess equations that are highly symmetric, and then work out their consequences to see if they can explain phenomena. Instead of going from the phenomena to find beautiful equations, we guessed beautiful equations and then figured out if they could possibly describe the world. And it works amazingly well.

I find good life advice in the way he uses symmetry to guess at the solution and then work backwards. Sometimes, if you're not getting where you want to go, try to reconsider where you're starting. Think about where you want to be and then work backwards.

SHIFT YOUR PERSPECTIVE
TWO DIFFERENT THINGS CAN BE TRUE AT ONCE

Keating: I find the writing in your book *A Beautiful Question* very evocative and almost poetic. And you quote both Walt Whitman and E. E. Cummings.

Wilczek: I like poetry. It is mind-expanding. And of course, there is something to be said about describing things in different ways. This can be very fruitful. There are different levels of description, and poetry can add. My point is, it is not either/or. It can be both. You can have different descriptions of the same object or the same phenomenon, and each are valid in their own terms. Each answers important questions, but they answer different questions. And if you try to use one description to address inappropriate questions, it can run out of steam or actually be wrong.

In quantum mechanics, we learned that there are complementary descriptions of physical objects, one of which is appropriate to asking questions about position and the other of which is appropriate to answering questions about momentum or, roughly speaking, velocity. We have this fantastic sort of movement back and forth between energy, mass, and frequency. We have beautiful equations that require particles to have zero mass and yet construct a world where things do have mass.

What Frank is saying here is extremely deep and applicable to human interactions—not in some woo-woo sense of "we are all quantum waves" or whatever, but in the sense that when two people hold strong and opposing opinions, it might be that they are both right, even though they can't see eye to eye. If you say an electron is a particle, you are right. If you say it's a wave, you're right. But a particle is very different from a wave.

So how can both be right? The answer is complementarity. It depends on perception, reference, frame, observer, and other factors. So when you're having a debate about an issue or trying to reach a compromise, remember that the word compromise and complementarity have the same root.

If we can agree to disagree about the nature of the building blocks of the universe, how much more can we agree to find value in the opposing viewpoint that we might think is wrong or misguided? If nature can be fundamentally indeterminate, is it possible that human opinions can be indeterminate—can be wrong and right at the same time? This concept Frank brought up reminds me always to perceive a debate opponent's perspective as they see it, and in so doing, get a deeper appreciation for what is actually true.

A NOBEL IDEA

ASYMPTOTIC FREEDOM: WHAT IT IS AND
WHY IT REVOLUTIONIZED PHYSICS

Wilczek has done some of the most theoretical work featured in this book. But it is also in the grand tradition of scientific inquiry that has been pursued since the ancient Greeks: the effort to subdivide the world ever more finely in order to understand what exists at its core. By the time he came around, scientists understood that atoms (Greek for "uncuttable") could be broken into protons, neutrons, and electrons. Physicists began a quest to see if quarks could also be separated into constituent parts. The short answer is that they can't really. (The amount of energy required to split a quark apart winds up creating an additional quark—think of it like a magnet with north and south poles; if you broke it in half, you would have two magnets with north and south poles.) Wilczek's contribution to the story is to figure out why. The answer has to do with a specific nature of quarks known as asymptotic freedom.

His discovery built a bridge to the theoretical physics of the previous decade—quantum electrodynamics, pioneered by Richard Feynman and Julian Schwinger—and provided a way to understand the inner structure of the proton. In our efforts to understand what happened during the first moments of the universe following the Big Bang, we can only observe aftershocks still measurable today. Understanding the inner substructure of matter, which is quarks, helps us do that.

DON'T FEAR FAILURE OR SUCCESS
JUST KEEP SHOWING UP

Keating: After you won the Nobel, did you feel like you had some kind of superpower?

Wilczek: I did have a long time to think about the "afterlife." I admire many people who won Nobel Prizes, and I looked at their lives after the Prize. Some had done better than others. To me, the success stories—such as Richard

Feynman, Chen Ning Yang, and T. D. Lee, for example—are people who took it in stride and went on writing papers. There are other people who got intimidated by the Prize. They thought that nothing they could do afterward would live up to what they had done before or to the prestige of the Prize.

I wanted to be in the first group, not the second. So I planned right away that after getting the Prize, I would write some papers, even if they were mediocre, just to have done it, just to break the ice. That is what I did. And I kept working. Fortunately, my style has always been to do something, try to make a basic contribution in it, and then move on to do something else. With that style, you are often going to fail, so I got used to failure. I wasn't afraid of failure, before or after the Nobel Prize, so I didn't get intimidated by it.

Don't be intimidated by your failures or your successes. He may have won this human-provided award, but he still recognizes how little we as human beings know, and that impels him to continue wanting to make great contributions long after, even if there are failures along the way. It's different than, say, Jeff Bezos retiring on a mega-yacht, because science is an "infinite game." Unlike chess, you can never win science. However, like chess, science is comprised of many finite games which are winnable, like tenure or the Nobel (in my case, losable). Ultimately, in the scientific community, you can never win. You might win a prize, but you can never beat Mother Nature. T. S. Eliot said, "The Nobel is a ticket to one's funeral. No one has ever done anything after he got it." Frank utterly demolishes that sentiment. That's a powerful motivating factor for people to continue on, and it's a symbol of their curiosity and relentless passion. We should all remember to make sure we are motivated by the right things.

PAY ATTENTION TO YOUR ATTENTION
BE INTENTIONAL ABOUT WHAT YOU FOCUS ON

Keating: What makes you decide that it's time to write another book or to pick up a new project? Is there a rubric, or is it just whatever happens to interest you?

Wilczek: I like to say my operating system is *Think, Play, Repeat*. I think about something. Then I play around with the idea. Then I think about it again and see if it gets better. The basic factors are the importance of a problem and my perception that I can do something about it.

But how do you decide what's important? Things can be important in different ways. A thing can be important if it's fundamental, meaning it's a basic principle about how the world works that cannot be explained in terms of more basic principles. A thing can be important if it has the possibility of doing useful things in the world—and we can allow a very broad definition of useful, such as pushing the frontiers of knowledge in some way or making something observable that was only latent about the world before. Importance can also come from just making things more beautiful, more aesthetic, and improving the description of the world.

And then, if I perceive flaws or gaps in our understanding of the world, that's another factor. You might jokingly say it is irritation. So, important, addressable, and irritating: those are the three axes.

When you see something in your life that you deem important, that you think you can change or affect, or that irritates you because of its incompleteness, pay attention. Those are key indicators of interest and eventual success. Further, it's important to cultivate the kind of situational awareness that leads us to notice key indicators. Or if Wilczek's rubric doesn't work for you, come up with your own way to determine what interests you and what you are curious about. It will likely lead you down more satisfying paths. Let your attention guide you. And, in valuing your attention, be choosy about where you give it—in some ways, attention is our most perishable resource.

STAY HUMBLE
REMEMBER—AND IMAGINE—HOW MUCH YOU DON'T KNOW

Keating: If complexity is an emergent property of the universe, do you believe there will be found evidence for extraterrestrial intelligence? And would they have structures like religion and civilizations, if they do exist?

Wilczek: There are circumstantial reasons to suspect that it should be abundant. And it's not a paradox that we haven't seen it, because it's far away and difficult to communicate. Here on Earth, as soon as the conditions became not crazy—as soon as the Earth began to solidify, cool down, and have a reasonable degree of stability and formed liquid oceans—it didn't take very long for life to emerge. The chemistry is complicated to get life started, but not crazily complicated. I believe it's very likely that life would start on a large number of the planets people are discovering now, extrasolar planets. They are not a rare thing. Zillions of [exoplanets] in the galaxy have many billions, probably, of potential [lifeforms]. And then, of course, there are lots and lots of galaxies.

I do think life is abundant in the universe. However, again consulting the history of life on Earth, it took a long time for multicellular life to emerge. It took special conditions: stability, having a nice star, a leisurely evolution. Then apply an orbit that's precisely circular, and a moon to kind of stir things up, and plate tectonics. Although we can't say what would've happened if any of these things hadn't been in place, it does seem that the step from a simple life is a difficult step that, on Earth, took a long, long time. It took billions of years. And then on top of that, to get from there to something we would recognize as intelligence, with use of language and abstract concepts—that's only been realized in humans, basically. And, of course, the growth of technology is only about two hundred years old.

I think extraterrestrial intelligence and extraterrestrial technology might be rare even though extraterrestrial life is common. I find it hard to get excited about a UFO. On the other hand, you know there's Arthur C. Clarke's third law, which says that "any sufficiently advanced technology is indistinguishable from magic." So who am I to say what some advanced civilization might be doing?

First, I love the serendipity of this Clarke quote, as it is the same one that introduces every episode of the "Into the Impossible" podcast. I also find great humility in this bit of our conversation. He recognizes how much he doesn't know and is unwilling to stake a claim on either side of the debate, instead turning his focus to what we do know, to whatever evidence can be used in comparison. I also appreciate and respect his willingness to discuss this topic at all, which some physicists dispel.

DON'T MINIMIZE WHAT YOU DON'T BELIEVE
IF IDEAS ARE IMPORTANT TO OTHER PEOPLE, WE CAN LEARN FROM THEM, PERIOD

Keating: Does your study and knowledge of religion satisfy something in you that your equations cannot? Or is it just a philosophy? What is religion to you?

Wilczek: I do not adhere to any established religion. I was fraught, growing up and into my teen years. I was absolutely fascinated by and deeply steeped in Roman Catholicism. A lot of my cultural heritage is from that. I can refer to it because it's part of my intellectual furniture. And I still have a lot of respect and admiration for that tradition. It's quite flawed. It's quite complicated and human. And it contains some very beautiful things.

The Judeo-Christian tradition is part of who we are, if one grew up as you and I did in the twentieth and twenty-first century United States. It's everywhere around us. To ignore it or minimize it, to attempt to stigmatize it—it's like cutting off your arm. If you do, you lose that cultural reference. There are things about getting around in the world and making sense of it that science doesn't answer. It can give you insight about the consequences of different choices and lead to a kind of wisdom at that level. But it can't ultimately decide what you want to do or what's right, what's good or what's bad, or what's in different categories.

So I think it's not a good idea to ignore or trivialize these traditions. In the spirit of complementarity, you can learn from them and also relate better to people who come from different perspectives, if you take it seriously and try to understand what they're thinking about.

To make our work as scientists easier, to isolate whatever we are studying, we tend to ignore whatever isn't part of or doesn't serve the experiment. Although this is incredibly helpful, it can also be dangerous. In Frank's above discussion, I see an unwillingness to ignore something that impacts humanity as much as religion does. I appreciate that dedication. It takes great effort to maintain a nuanced approach to life and work. But it's very important to.

INSIDE A NOBEL MIND
FINDING THE HUMAN IN THE GENIUS

Do you ever fall victim to the imposter syndrome?

Wilczek: I really don't. And there is a reason for it, which is very concrete. I got over the imposter syndrome when I went to college and maybe even before in the New York City school system because we had a lot of tests, grading, and tracking. The consequence of early success is confidence. And that's extremely important to a research scientist. Confidence is pure gold. It allows you to think in different ways, not to worry about what other people think so much, and to believe you can take on big problems because you think you have a chance of doing better than other people who have done them.

What do you leave in your ethical will, to bequeath to humanity?

Wilczek: I have my body of work, but I'm actually starting to have what will be a broader cultural legacy. I want to enrich my family's lives and [the lives] of the wide body of friends all over the world, including institutions in Shanghai and Stockholm that I have helped develop. So I'd like to see those thrive. I'm writing mystery novels. I'm thinking about putting together my Wall Street Journal columns that go off in different directions, opening up to cultural and scientific applications. And if the opportunity arises, in connection with climate change, control of nuclear weapons, or addressing the challenges of artificial intelligence, I'd be happy to contribute some wisdom to those questions.

If you had a monolith that would last a billion years, what would you put on it?

Wilczek: Our Core Theory [Frank's name for the Standard Model describing the known laws of nature] is something you can encapsulate in a short computer program. So I would put that on a USB stick.

If you were able to time travel and give young Frank Wilczek a piece of advice, to give him the courage to go into the impossible, what would it be?

Wilczek: You shouldn't feel guilty about exploring, investing in exploration, as opposed to trying to immediately exploit the things you've learned and the skills you have. Explore before you dig or while you're digging. There's plenty of time to think a lot of things.

KEY TAKEAWAYS

- What an amazing thing it is to have the kind of patience Wilczek exhibits. We can all learn from and apply this equanimous approach to life. Whether waiting for a child to be born, for a wedding day, or to get a PhD in graduate school, we have to enjoy the process. The journey is its own reward, not the gilded destination that might or might not wait at the end.

- People often marvel over how young Wilczek was when he did his Nobel Prize-winning work. While this is true, the underlying lesson here has more to do with accumulated knowledge. He was guided not only by his intuition but also by careful analysis of what was already known not to work. Physicists benefit from accumulated knowledge, even including wrong turns. Success is ultimately cumulative. So is wisdom. I marvel even more at how much Wilczek continues to add to our collective knowledge since his Nobel Prize-winning work.

- Wilczek teaches me that we have to look for ways to find our confidence—and remember that it comes from a variety of places. When I talk to my students, I point out that there is an importance in the grand slam but there's also importance in the singles that get points on the board. When we remember to value those individual contributions, we can gain confidence from them and, hopefully, like Frank, shake the imposter syndrome at the same time.

- I also respect that he doesn't shy away from discussing God and spirituality, topics on which many scientists demur. He is willing to and seems to enjoy talking about the big picture, even soliciting new-age gurus like Deepak Chopra to endorse his books. I think that makes him a very deep, very accepting person, and also makes him and his work relatable. I felt like he could have and would have talked to me for hours, and he brought an incredible amount of what I call soulfulness (even though he would probably disagree with that word!) to our conversations. Frank Wilczek is that rare breed whose EQ matches his IQ. His vulnerability is as refreshing as his sheer intellectual prowess.

CHAPTER 8

JOHN MATHER: THE COLLABORATOR

John Mather is a senior astrophysicist at NASA's Goddard Space Flight Center and a professor of physics at the University of Maryland's College of Computer, Mathematical, and Natural Sciences. In 2006, he and George Smoot were awarded the Nobel Prize in Physics "for their discovery of the blackbody form and anisotropy of the cosmic micro-

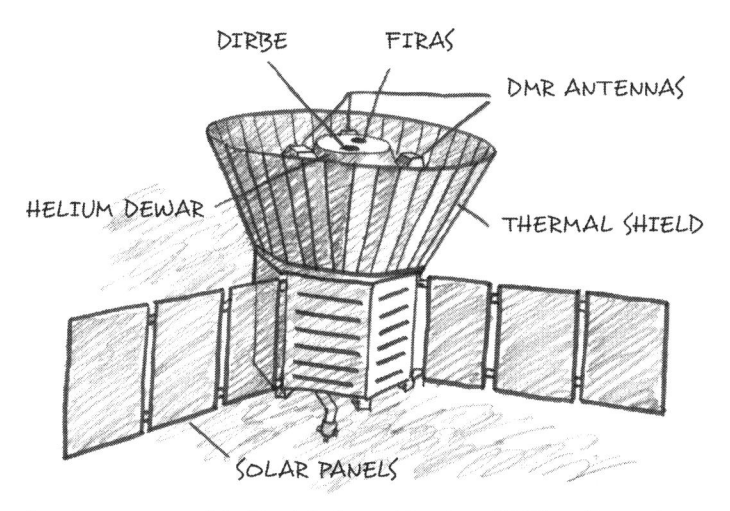

The major components of the COsmic Background Experiment (COBE) satellite are shown here, in particular, COBE's Far Infrared Absolute Spectrophotometer (FIRAS), which firmly established that the Cosmic Microwave Background was the byproduct of an extremely hot and dense phase in the early universe, when the entire cosmos was essentially a fusion reactor.

wave background radiation." The work, completed along with their teams, used the COBE satellite to all but confirm the Big Bang theory—and elevated cosmologists, individuals previously derided by "real" physicists as being "always in error, but never in doubt," into proper practitioners of precision science.

When I started off as a young student in the early 1990s, I was almost ashamed to admit I was learning to be a cosmologist because there were important attributes of the universe that we were woefully ignorant of. Mather's work changed that. For example, back then, we didn't know if the age of the universe was 10 billion years or 20 billion

JOHN MATHER: THE COLLABORATOR · 135

years. That's like looking at an adult and not knowing if they're teenaged or elderly! There were stars in the Milky Way galaxy which we knew were older than the reported age of the universe itself. That would be like a child who is older than their parents. It was laughable! But the work of John and his team in the early '90s revealed that we could know with great precision quantities that were previously ambiguous. That resolution of ambiguity gave me great clarity that this was the field I wanted to pursue.

John is another of those laureates who puts to rest the canard that winning the Prize is the end of a career. John has gone onto tremendous success, both after the discoveries made by COBE in the 1990s and also after winning the Nobel Prize in 2006. He is currently the project scientist for the James Webb Space Telescope, a multi-decade and multibillion-dollar project aiming to succeed and enhance findings gathered by the Hubble Space Telescope. He's moved into a completely different field from the cosmic microwave background, showing that he has an extreme depth *and* breadth of knowledge. Because he's playing as significant a role on the James Webb Space Telescope as he played with COBE, I expect great success to come from the project.

DON'T FOCUS ON THE OUTCOME
IF YOU DON'T *THINK* SOMETHING IS IMPOSSIBLE, THEN IT MIGHT NOT BE

Keating: Do you have any anecdotes from a mentor to share?

Mather: One thing [my PhD thesis adviser Paul Richards] said that still sticks with me is, "In your talks, people can't absorb everything you're going to say. So remember the three things you really want people to understand and make sure they get that part."

Keating: What advice from a mentor has most affected you?

Mather: To [successfully mentor people], give them really hard problems, maybe impossible problems. Don't tell them that they're impossible, and let them figure them out. And so I have taken on really hard problems and they've gradually yielded bit by bit so that we could actually build a COBE satellite and even the James Webb Space Telescope. So our whole community has been doing this bit by bit, and we've been inching forward to make better equipment to make our measurements that would have been completely imaginary.

Often, when we are working on something, the task seems impossible. Perhaps that's because it is. Or, you might have that feeling because you're on the precipice of a breakthrough. If you don't persist, failure is all but guaranteed. Everyone who has succeeded had to persist. Whether or not something is impossible can't be determined if you give up too soon.

A NOBEL IDEA
THE COBE SATELLITE: WHAT IT IS AND WHY IT REVOLUTIONIZED PHYSICS

John Mather, along with his fellow Nobel laureate George Smoot and their team, was a driving force behind the COBE satellite and its monumental discoveries. The satellite combines the power of extreme sensitivity with clean, high-fidelity measurements that are almost as free of imperfection as an instrument can be. Through its measurements, the team proved, unambiguously, that our universe was once a nuclear fusion reactor, producing the very lightest elements on the periodic table, which would themselves then be used as fuel to create stars, which then would produce planets and people.

John and his team revealed with exquisite precision the underlying properties of the nucleus of the nuclear reactor that was our baby cosmos. This not only allowed us to understand the precise composition of the universe but also effectively killed off competing theories that did not invoke a Big Bang, such as the so-called Steady State Universe. After John's work, only a tiny fringe group would maintain the existence of anything other than the Big Bang that John and his teammates ushered in.

CONSIDER THAT YOU'RE WRONG
BUT DON'T BE NEUROTIC AND WASTE TIME

Keating: What do you say to people who are so bright but nonetheless believe heterodoxy?

Mather: I usually don't think it's worth arguing because people who have made up their minds have their own reasons. It would be interesting to ask, "Well, why do you still believe this when we have this other evidence?" But I have not made that question. Looking at it from the outside, we say, "Well, how can you possibly imagine that your story produces this degree of perfection? As far as I know, they can't do it. But I guess they think they can!"

Keating: Do you think we can ever have unanimity in science? And do you think it's good or bad to strive to achieve unity, when maybe we can't achieve it?

Mather: I don't think it's an objective we should [keep] secret. Nothing would make people happier than to discover that we've all been wrong about something. Then we've all got something to do, and we've made an advance of some sort.

Keating: Richard Feynman said, "Science is the belief in the *ignorance* of experts," meaning, for example, that if Einstein had just trusted Newton, we would never have heard of general relativity. I don't know a scientist who says, "Oh, so-and-so is such an eminent scientist, I'm just going to believe whatever they say."

Mather: Yeah, that's not us. We don't do that. The general public might think we're all group thinkers because we all say the same thing, but that's because the evidence is pretty strong. So I don't think we're stuck in some kind of groupthink the way some of my colleagues think. I think we are very busy pursuing the possibility that we are all wrong. But each scientist has to make some judgment about where the payoff is likely to be. So if you're looking where it's pretty unlikely that you're going to find a discovery, then, well, after a while, you might get tired.

I've always felt that even when a development has become accepted as orthodoxy, the act of engaging with alternate theories can at least serve to strengthen the orthodoxy. So I think there's value in heterodoxy, to help us avoid the perils of groupthink. Sure, as Mather suggests, too much of it can be a bad thing, when engaging heterodoxy winds up wasting time and distracting you from continued work (and finding this balance is especially important as you gain more success, since the higher you fly, the more people will want to take shots at you). But it's also important to remember that heterodox views sometimes lead to breakthroughs. When Lemaître first posited the idea of the Big Bang in 1927, it was heterodoxy. The key is to have enough of a crack in your monolith of orthodoxy to let a little light in. Since it takes a lot of power and time to topple a monolith, there's no danger in a small crack. The other option—being so cautious that you allow no light to enter any cracks—means you never grow, adapt, or change.

DEFEND WITHOUT BEING DEFENSIVE
KEEP IT ABOUT THE WORK

Keating: In science, it can take years before work is confirmed and a consensus is built. During that anxious time, how do you handle the attacks on your ideas?

Mather: There are lots of different jobs in science, and my job, as I looked at it, was to build some equipment and go measure something, and then the interpretation [of that evidence] is some other person's job. So when I say, "This is what we found," I think it's right and we've done the best we can. We've got a big team of scientists. We all looked backwards and forwards all over everything—we couldn't find anything wrong that we did, we haven't thought of anything wrong that anybody else did, and we haven't thought of a way the universe can fool us. That's our job. And then plenty of other people say, "Well, no, that can't be right." When we showed our spectrum at the astronomical society and got a standing ovation, somebody heard the guy cheering the session say, "They've swallowed it, hook, line, and sinker."

In the end, John's critics proved incapable of the task. He was resilient in the face of criticism, turning again and again to the evidence. It's important to listen and respond to your critics, but it's just as important not to internalize them. Sometimes people in your field will be more interested in glorifying their own egos than sharing what should be the common goal of advancing human knowledge (or whatever your field defines as success). Although you can't ignore such detractors—you should always defend your work and invite collaboration—you can focus on resilience and patience and let history be the judge.

ACCEPT WHAT YOU CAN'T DO
AND USE YOUR TEAM TO FILL IN THE GAPS

Keating: I want to ask about teams. How do you hire people as a lead scientist on a project? How do you think about the work from a team perspective? How do you manage and build a team and inculcate it with values?

Mather: I wish I could tell you! When I got to Goddard Space Flight Center, I was only thirty years old. I said, "I do not know how to do any of this! I'm gonna go to my meetings with my fellow scientists and engineers, and we'll work together on this, but I'm not the person to manage this project. Don't even ask me." I had no idea you had to do that. So Goddard assigned professional scientists and engineers, who were my supervisors and my mentors, and they did the really did the hard job of figuring out how to make this thing work. So we were able to join an existing engineering organization that basically already had all of the talent that was needed. They knew how to do this stuff that people think I did.

This is such an important lesson to share. Note that this Nobel laureate has the humility to share it! He didn't know how to manage, so he made sure the right experts were hired and he listened to them. He was a servant leader in that sense. He put aside his pride and made sure they got the best people required for the job. In pro sports, managers need to get the best players on their teams, not be the best players on their teams.

BEND WITH THE WIND
BE READY TO PIVOT AT THE FIRST SIGN OF A STORM

Keating: After the Space Shuttle Challenger disaster in 1986, you almost had to start from scratch with your plans for COBE. How do you overcome those kinds of major challenges?

Mather: I don't think there's a big plan—you deal with them when they arrive. Some of our managers said, "We'll find a way. We didn't get this far to give up." That's how everybody felt. So, "Let's get a different rocket." It was quite a significant accomplishment to find one when they weren't available, but we did. As it turned out, it had to be assembled from parts that had been lying around because there was not actually a whole rocket available. But people knew that the project we were building was important enough to rescue. People realized that if we could solve this problem, the COBE satellite could be the first thing we [at NASA] launched for science after the Challenger.

Luck works both ways. Sometimes you'll get undeserved lucky breaks, as John described earlier. Other times, luck will reach up and smack you in the face. The Challenger disaster was a tragedy, not only for the families of the astronauts who were lost, for science, and for the country itself, but, of course, also for NASA. The ramifications of that could have been the end of COBE. That same space shuttle was set to launch COBE the very next year. Even so, the entire shuttle launch program was grounded and delayed in a safety stand-down, in part presided over by physics Nobelist Richard Feynman.

Fortunately, the program continued, and NASA emerged stronger and more resilient than it had been before. Ultimately, because of the team's perseverance, they became more disciplined and were able to make the best of an awful circumstance. Mather's answer to this question makes me think of the proverb attributed to that wise scholar Confucius that tells us "the green reed which bends in the wind is stronger than the mighty oak which breaks in a storm." Might is a virtue, but there will be things completely out of your control. It's inevitable that you'll face headwinds. Be flexible. Be resilient. Storms pass.

COOPERATE TO COMPETE
THE GOAL IS SO MUCH BIGGER THAN JUST WINNING

Keating: There is a tendency, in popular imagination, to think that science is about credit, egos, and politics. In your outreach to the public, how do you get beyond that?

Mather: My perspective is that we are in this together. Even if we are competitors. If you're working on a project and I'm working on a project that's supposed to measure the same thing, and we don't get the same answer, that's really important. So our job is to get evidence, not to win. If we can get evidence, then that is a win! I think a good scientist is always saying, "I'm going to get more evidence. I'm not trying to browbeat anybody into believing me. I'm just going to get more evidence." I never worried much about credit. The only way to get credit is to give credit. If you think you can capture credit by claiming you did something, well, that lasts for a week, and afterwards, people will remember that you're the one who took their credit away from them. The truth is that we are all in this together, even if it doesn't feel like it.

There's as much competition in science for prestige and credit as there is in any Fortune 500 boardroom. John reminds that we are looking ultimately for truth. Although it's okay to be competitive, you can't let the competition get in the way of the pursuit of truth. This applies both to scientists, who ultimately have the same goal, regardless of the small gains they might make along the way, and also to, for example, the car salesman, who might make more sales in the short term by playing politics and competing for credit but will make far more sales in the long term through word-of-mouth referrals and by having integrity and honoring truth—even when that means sending a customer to a competitor.

INSIDE A NOBEL MIND
FINDING THE HUMAN IN THE GENIUS

Do you ever wrestle with the imposter syndrome?

Mather: Every day I get up and work on something I can't figure out yet. So every day I'm the imposter who doesn't know the answer. Now what am I going to do? I don't have any choice. This is my job. I have to work on the thing I haven't figured out yet. Once in a while, I say, "There's somebody over there that's really smart." What do I do? Ask that person for help. Don't try to beat them; try to join them.

As you know, Alfred Nobel's prize money goes to a person who discovered something that not only is interesting but also benefits mankind. In that sense, he left a will that is both material and ethical. What ethical wisdom do you want to leave for your ideological heirs?

Mather: That's a big challenge. I think we can't really see the grand picture, but we can see the local picture, and we hope that it propagates. We can say, "I'm going to do the best I can with the people I am with," so we all proceed ethically together. We do not cut corners. It's pretty clear that if our society is to survive for more than a few centuries, we're going to have to place immense faith in the work of scientists and engineers. So I do what I can in that direction.

What piece of information would you put in a time capsule guaranteed to endure for millions or billions of years?

Mather: What I'd like to get on the next Voyager record that goes out of the solar system is the United Nations' Universal Declaration of Human Rights.* It's a way of saying we are all in this together. And we do actually intend to treat ourselves and each other with respect and dignity. I think we could make it happen, but we do need to understand ourselves a little better than we do.

What would you tell yourself as a twenty-year-old kid that would give you the courage and benefit of wisdom to go into the impossible as you have?

Mather: We are all in the same position that we do not know the future. We have the opportunity to look for things to do that nobody's ever done. That's a scientist's job. Maybe it's everyone's job! Once in a while, an idea comes to mind, and I say, "That would be really cool if we could do that!" And usually it's impossible or nearly impossible. The mindset that makes that interesting is, "Well, why not? Let's try it and see what happens."

---

* https://www.un.org/en/about-us/universal-declaration-of-human-rights

KEY TAKEAWAYS

- Don't be motivated by credit. Instead, set your sights on collaboration. Harry S. Truman—who may not have won a Nobel Prize but *was* a two-term president—famously said, "It's amazing what you can accomplish if you do not care who gets the credit." Mather feels the same way.

- Mather resonates with me because he exemplifies the power of local thinking. He says his job is to get evidence, not proof. He emphasizes building a team to cover all the required skill sets. He says we can each only see the local picture and hope that it propagates outward so that we can discern the full majesty of the tapestry. If everyone does their part and doesn't cut corners or undercut each other, then everything will get done and the universe is richer. In that way, it's equally important to focus on the implications of what you do each day and how that work contributes to the ultimate symphony. Only when each musician in the symphony comes together do we witness the true purpose of the endeavor.

- Don't get caught up in worries over whether the outcome of your work is possible or impossible. Thinking something is impossible could just be a limiting belief. As Audrey Hepburn said, "Nothing is impossible, the word itself says 'I'm possible'!" Just take one step at a time and see where you get. Problems that John thought were impossible gradually yielded bit by bit. Further, if you're taking one step at a time, you'll be able to quickly pivot your path when a challenge appears on it.

CHAPTER 9

BARRY BARISH: THE AVUNCULAR AVATAR

Barry Barish is the Linde Professor of Physics, Emeritus, at Caltech and faculty member at UC Riverside. He became director of the LIGO (Laser Interferometer Gravitational-Wave Observatory) project in 1997. In 2017, he was awarded the Nobel Prize in Physics along with Rai Weiss and Kip Thorne and their teams "for decisive contributions to the

Two black holes, each about 30 times the mass of our sun, spiraled together some 1.3 billion light years from earth and collided, producing an eruption of gravitational-wave energy finally detected by the LIGO experiment in September 2015.

LIGO detector and the observation of gravitational waves." Before joining the LIGO experiment, he worked on the Superconducting Super Collider, the high-profile particle accelerator that was canceled by Congress in 1993. He has many other awards, is a member of the National Academy of Sciences, and served as president of the American Physical Society in 2011.

Barry is a consummate scientist. He has hands-on technical expertise, the interpersonal skills to motivate and lead, and the scientific discipline to know when to quit and went to double down. He's an expert at the kind of soft skills scientists typically don't pay attention to: relationships, networking, and the importance of mentorship and communication. And all of this is fueled by curiosity—after our interview, he asked if he could interview me because he was curious about some challenges I discussed in my first book, *Losing the Nobel Prize*. That conversation was one of the greatest thrills of my life. Barry, at his core, is truth. He is not afraid to be vulnerable, as you'll see. He's the type of scientist—the type of human being—I want to be because he is generous, gracious, insightful, guileless, and avuncular. I hope to cultivate these traits not only in myself, but also in my students who trust me to guide their careers.

DON'T GET TOO COMFORTABLE
DISCOMFORT LEADS TO GROWTH

Keating: You left particle physics when you moved over to the LIGO experiment. Was it scary to leave the field you had come of age in?

Barish: Invigorating. Why would it be scary? The only reason things get scary is if you are so comfortable in what you are doing, which probably means you aren't pushing yourself or doing anything very interesting. I have always done different things.

His courage comes through here. This stopped me dead in my tracks. It flipped things around for me, to hear that growth is painful sometimes but also leads to new synergies that can catalyze discoveries. You'll bring to your new endeavors the previous tools developed in different fields. That can be a pow-

erful combination: the freedom of having a beginner's mind combined with the advanced tool kit from decades of prior experience. It's almost welcomed to take on challenges like this because on the other side of fear is stagnation or worse. Such challenges are growth opportunities. Seeing them that way is, yes, invigorating. He's helped me develop the mental toughness to lean into these kinds of changes, to go where the fear is.

FORGET GLORY
FOCUS TOO MUCH ON PRAISE AND YOU MIGHT NEVER GET IT

Keating: When the Superconducting Super Collider was canceled, it freed you up to work on LIGO. Otherwise, LIGO might not have happened, and you wouldn't have won a Nobel Prize. Looking back, could the cancellation of the Super Collider serendipitously have been one of the best things to happen to you?

Barish: What you say is true in the sense that, I like to think, we detected gravitational waves when we did partially because of myself. The [decisions that enabled us] are not all due to me, but some of them are.

Barry dodged the question about the Nobel. I'm including the exchange anyway to highlight the fact that he was never driven by the Prize. He's a bigger man than I am! With a few exceptions, most laureates share this indifference to adulation. I find this remarkable, perhaps because of my own desires to win it (and probably this means I never will). When you are indifferent to adulation, that frees you to focus more on the process, to do the work that would lead to adulation anyway.

> **A NOBEL IDEA**
>
> **DETECTION OF GRAVITATIONAL WAVES: WHAT IT IS AND WHY IT REVOLUTIONIZED PHYSICS**
>
> Kip Thorne, Barry Barish, and Rainer Weiss, together with their team, built, developed, and delivered the scientific results from the LIGO experiment, which resulted in humanity's first ever direct detection of gravitational waves, which are slight ripples in the curvature of space-time that move at the speed of light. These detections have revolutionized our understanding of so-called compact objects, such as black holes and neutron stars, as well as the role they play in the astrophysical processes of our universe. It is a discovery on par with Galileo's discovery of craters on the moon, marking a watershed moment in humanity's history, not just in astronomy.
>
> The project is the combination not only of extremely powerful technologies—such as laser and vacuum technologies—but also of disciplines. It required very sensitive and close interaction between theoretical physicists and experimental physicists. That union resulted in the spectacular discovery, which in fact was expected: Einstein predicted waves of gravity back in 1916. However, he

thought they could never be detected because of the incredible technological challenges of the last century. For decades, we only had circumstantial evidence. To persevere, as the LIGO team did, for forty years—raising funds, raising awareness, and keeping the faith that these theoretical waves could be measured and "seen"—is itself a spectacular achievement.

TEACH AN OLD DOG NEW TRICKS
BRING A NEW APPROACH TO EXISTING TOOLS TO SOLVE NEW PROBLEMS

Keating: What is the most fascinating, intriguing aspect of your job on LIGO?

Barish: I always thought that it represented a new way to look at the sky, using gravity instead of photons. That meant we could look at the sky to do a new kind of astronomy. Because gravitational waves aren't absorbed, if we can see them directly from the early universe, we can see back to the first instance of the early universe, instead of a few hundred thousand years after. It's just a new approach. I was romantically attracted in these kinds of ways.

This speaks to the power of bringing a new perspective to existing tools. Conceptually, LIGO is no different than other types of observatories in that it uses signals that travel at the speed of light to inform us about the most distant reaches of the universe. However, it's a completely distinct technology and field of study.

The analogous transformation in the field of astronomy is what Galileo did in 1609. He was the first person to use a telescope to look at astronomical objects and in so doing revolutionized our understanding of the entire universe and our place within it. Since then, we've used them to study the origin of galaxies, the origin of stars, and the beginning of time. Likewise, we don't yet know the untold riches of discoveries that LIGO will lead to. Sometimes, we can use old tools to solve new problems simply by approaching them in a different way.

DON'T MISUNDERSTAND THE WORD IMPOSSIBLE
SOME THINGS CAN'T BE KNOWN; OTHERS JUST CAN'T BE KNOWN YET

Keating: Who says there has to be a unification of quantum mechanics and gravity, a so-called theory of everything? Maybe we are just prejudiced to try because Einstein and others have tried.

Barish: I think you are right. As scientists, it is attractive to think there must be a bridge between them. I think that is worth pursuing. Can you really say that that absolutely has to be the path to truth? I don't think so. It may be that that is a false direction. But searching for gravitational waves or cosmic microwave background—all these things that we do that haven't been seen before—we would not be looking if we knew absolutely that the direction is right. So it has to be promising enough: promising in that it might be the right answer and promising in that you can actually make progress.

Barish and his team had incredible courage to attempt things that seemed impossible. They wanted to find the limits of what was possible to do, so they had to go into the impossible. In Galileo's day, the prevailing wisdom held the moon to be a smooth ball of perfect crystalline material. It was impossible to conceive of it otherwise. But did that absolutely have to be the truth? No. Galileo had courage to imagine the seemingly impossible. Some things are incredibly interesting but impossible, like entering the event horizon of a black hole and living to tell the tale of what you saw. But some things only seem impossible and in reality are just technologically challenging. LIGO sought to measure something in a way that seemed like it would require magic. But the team was not deterred. As Arthur C. Clarke said, "The only way of discovering the limits of the possible is to venture a little way past them into the impossible." That's what Barry and his team embodied.

LOVE THE ONE YOU'RE WITH
DON'T TRADE IN TOOLS THAT STILL SERVE YOU

Keating: Is the Hubble tension of interest to you personally?

Barish: We measured the Hubble constant, actually. If this remains a tension a decade from now, I think we'll resolve it. But we need a thousand times more data to pursue that.

I don't think we are limited by systematics, just limited by getting enough data. I would say we are a decade away. But it's doable in LIGO, not a future instrument. It will be done with the improvements we will make in LIGO.

Yes, they would have to improve the instrument significantly to achieve this kind of measurement, but he's saying we don't need a brand new billion-dollar instrument—which may not even get us what we want. Make use of the tools you have in hand. Extract all the juice out of your current situation before hitching yourself to a future idealized wagon that might not even carry you where you want to go. Make the most of what you have before switching to the next shiny thing.

TAKE BIGGER RISKS
DON'T KNOCK AN IDEA BEFORE YOU'VE TRIED IT OUT

Keating: When do you cut off an experiment? When is it time to move on?

Barish: The general problem we have—and why our kind of science doesn't move forward faster than it does—is that the system is too conservative. We love something called peer review. But peer review is actually very conservative. We turn in proposals to get funded. If our proposal is offbeat, it doesn't get all "outstandings" in the reviews. And my money came from the NSF, which is answerable to Congress. We have to be able to tolerate failures, many more failures than we do. Experimentalists should be turned loose to follow our dreams, and I think science would move forward much more.

Big rewards require big risks. This speaks to a common theme in this book about the importance of "useless" research. When it works out, we make huge discoveries: for example, black holes colliding together or supernovas blowing up in far distant galaxies. Sometimes, being too conservative at the beginning compromises the high rewards to come at the end.

DEVELOP BESPOKE SYSTEMS
ONE SIZE DOES NOT FIT ALL

Keating: As a manager, you're very deliberate about who you hire and for what roles. I've asked you before for the shortcut advice on management—books to read, courses to take, gurus to follow—and you always say, "Sorry Brian, there is no shortcut." How did you develop your science-research managerial skills?

Barish: First, management has to be done by scientists because there are too many scientific decisions. Having a bunch of managers doesn't work. But the key to success is integration. For example, on LIGO, we have the world's best laser people, and a fantastic group that does controls, and so forth. How do you bring them all together to sing as one instrument? And then you have to do something to kill or mute the change-control system that exists in an organization, because that inhibits change, and you want to be as current and forward-looking as possible at all times.

This floored me. Before determining all of this, Barry went out and studied traditional organizational structures and managerial systems, determined what would and wouldn't work for his own circumstances—in academia, which is nontraditional because it lacks the hierarchical structure of, for example, industry or finance—and then essentially reverse-engineered a management style specifically designed for his own project. One size does not fit all. Imagine if every organization in every industry "read the manual"—and then threw it out and wrote their own. Systems would run much more efficiently and successfully.

INSIDE A NOBEL MIND

FINDING THE HUMAN IN THE GENIUS

What wisdom would you like to leave for humanity?

Barish: I am going to be very practical. Young kids—five years old, seven years old—are incredibly curious. They want to know everything. Then we get kids that come to Caltech and they don't ask questions anymore. They just want to do their homework. They'll ask you a question only because they couldn't solve a problem.

Somehow, in the educational system, we kill curiosity. We even have a saying which is detrimental: "curiosity killed the cat." That actually tells you that you don't want to try anything because it is going to cause trouble. That is the message: to anyone who is curious, do it; you are not going to get killed.

What was intimidating to you as a young person and seemed impossible but, through courage or hard work, became possible? What advice would you give your former self about that?

Barish: The big problem I had, although it doesn't show so much now, is being overly shy—shy in every aspect of being a person. I may in my head have been adventurous, but I was very reticent until I got enough confidence through success that emerged later in life. I am lucky that I went down a path where I succeeded enough to develop inner confidence.

Do you suffer from the imposter syndrome?

Barish: I do. I will tell you an anecdote from the Nobel ceremony. Mostly, it's intimidating to have this king give you this thing. But then they walk you up at the very end to the Nobel Foundation, they take official pictures of you, and they hand you this little book. It is leather bound and so forth. They open it up to a page and ask you to sign your name. I'm curious, right? So I look back on the previous pages and there is Einstein's signature and Richard Feynman's. How do I belong in this same book? I certainly have had the imposter syndrome, and I had it dramatically at that moment.

KEY TAKEAWAYS

- Imagine that your whole life has culminated in building what the world has decided is the pinnacle of what physicists have ever made in the history of humanity: the Superconducting Super Collider. And then it's canceled for craven political reasons completely out of your control. Most people would view this as a horrendous setback. When Barry faced that situation, he just walked away and said, "Okay, where do I go next?" Further, although he had the opportunity to join a similar project—the Large Hadron Collider in Europe, work that closely aligned with the skills he'd built for decades and that eventually succeeded in measuring the Higgs boson, the so-called God Particle—Barry instead made a huge, risky leap into a completely different field to work on LIGO, which I find so courageous. He turned the tragedy into a literally golden opportunity.

- From Barry, I learned about the importance of breaking molds—or at least not being bound by them. He wasn't constrained by the way his past tools had succeeded. He designed a new managerial system specifically to suit his needs rather than blindly adopt an existing system. He didn't even stick to the safe, conventional mold of being a particle physicist. He reminds me to see everything I approach with fresh eyes, including myself.

- This interview made me think about the dangers of stifling ourselves with limiting beliefs. We dampen our curiosity as we age. We minimize our ability to succeed by second-guessing whether or not we belong. And we hamstring the achievement of big rewards when we think too conservatively. Barry speaks of how important it is to encourage curiosity, confidence, and risk-taking, and also therein suggests that creativity itself is what we'll otherwise lose.

CONCLUSION

There's an old joke that goes like this: How do you know when a scientist is an extrovert? When they talk to you, they look at your shoes. I've heard this joke many times, and told it myself more than I should have. But it's not necessarily true. With this book, I wanted to demystify scientists. That was a goal from the start, because the more scientists are held up as unapproachable, inaccessible geniuses, the fewer young scientists—especially women, minorities, and marginalized people—will enter the field. As a result, we will miss out on the next true revolutionary.

When I sought to demystify the subjects of this book, though, I also was fearful. I thought I might face an insurmountable challenge. I assumed these medaled scientists were largely the beneficiaries of privilege and luck. Many in this book were fortunate to go to elite high schools. Further, once their careers began, they more often than not found themselves in the right place at the right time, enabling them to join a project bound for Nobel glory. If that were the case, then what was there to learn from them? Telling people how to think like a Nobel Prize winner would be as useless as giving advice on how to win the lottery.

But they all admit the role that luck played, while simultaneously exhibiting the kinds of work ethic and determination that prove luck is never enough. As the adage goes, fortune favors the prepared mind. While the public knows and occasionally envies these scientists, many among us still have not had the opportunity to appreciate the struggles they faced and the hard work they put in.

These laureates also falsify their fellow Laureate T.S. Eliot's claim that the Nobel is "a ticket to one's funeral...no one has ever done anything after he got it." That these individuals continue to strive to conquer new intellectual frontiers, using curiosity as their guide, is an inspiration to all of us: no matter what you have already achieved, your best days lie ahead.

Further, they discuss their work and successes with a lack of flamboyance, arrogance, or swagger. Their modesty and humility encourage the rest of us that what matters most is activity of mind and hustle. Although we can't emulate luck, we can match work ethic and philosophy. In these pages, they also reveal the secret sauce that fuels it all: curiosity.

PASSION VERSUS CURIOSITY

I find the common advice to "follow your passion" inert. When someone advises that, I almost wonder if they're only validating whatever your passion is as a way to stop the conversation. It's dismissive—and not necessarily good advice. It suggests you turn your avocation into a vocation. But loving a subject matter isn't enough to sustain a career in it. "Follow your curiosity" is much more sage advice. That's what I heard these laureates say again and again.

When curiosity drives you, you'll always go deeper than a hobby could push you to. When you're passionate about something, you engage in that subject matter to get a quick dopamine hit. But when you're truly curious, that dopamine hit would never be enough. Curiosity triggers different reward mechanisms that are more sustainable, that lead to resiliency. It's great to be passionate about things. You may feel passionate about more than a dozen things. But when building a career and legacy, put your time and effort instead into what makes you most curious. Further, if you can maintain curiosity—which is essentially an admission that there's more for you to learn about a subject—then you will never fall into the trap of considering yourself an expert. And if you don't see yourself as an expert, then you're less likely to suffer from the imposter syndrome.

CRUSH THE IMPOSTER SYNDROME

This project unintentionally became a self-help book for somewhat technically minded but mostly curious people who will want to magnify their wonder through the same high-level, very creative processes embodied by the laureates in this book. Science is creative. But there is creativity in every profession. If you feel there's no creativity in your work, that is itself a form of the imposter syndrome. And it's limiting you. Hearing these laureates discuss the importance of creative skills, such as communication and leadership, is a reminder to all of us that even if you are at the top of your field, if you don't have the soft skills to express yourself, then no one will know or understand your work.

The more I encountered these individuals as human beings, the more I saw they were ordinary people, facing the same insecurities, challenges, struggles, and fears we all face. A very powerful takeaway emerged for me only toward the end of my series of interviews, which was the recurring thread of feelings of inadequacy. Most of the individuals in this book do seem to suffer from the imposter syndrome—even after winning a Nobel Prize.

The imposter syndrome is a form of limiting belief. They say hatred is like taking poison and expecting it to kill the other guy—the imposter syndrome is like taking poison and expecting it to heal you. It's such a destructive bias. This book helped me find a way to break free of it.

I discovered a paradox: In the beginning of your career, nobody notices you, much less thinks you're a charlatan. Therefore, in the beginning of your career, take advantage of that anonymity to master your craft. Then, once you have acquired the skills to be a master, remind yourself gently and often that everybody has the same self-doubts and destructive inner narrative. You must give yourself enough compassion to grow beyond these limiting beliefs, because the only person who thinks of you as an imposter is you—not your college professor, your peers, or your boss, and certainly not the Nobel laureates in this book who almost all suffer from the same condition. That gives me great comfort, and I hope it does for you too.

ACKNOWLEDGMENTS

First and foremost, I'd like to acknowledge the nine men who shared so graciously their time, wisdom, and experience with me so that I might share it with the world: Adam Riess, Barry Barish, Sir Roger Penrose, Carl Wieman, Duncan Haldane, John Mather, Rai Weiss, Sheldon Glashow, and Frank Wilczek.

Barry Barish cast the die of this book in late 2020 when we sat down for an interview that would change my life forever. Spending time with this great man, a man about whom previously I really only knew legends, was a gift I will forever treasure. To call him a friend and have him write a Foreword to this book is truly surreal. I appreciate it more than I can communicate.

Yoni Falkson and Melissa Miller helped to organize nascent thoughts and brought out the juiciest nuggets that needed to be revealed.

Jane Borden was part editor, part sounding board, and all therapist. She kept me going when I almost gave up, more than once. She is a consummate professional and deserves a Nobel in patience, project management, and diligence. My deepest hope is that I can continue to create new works like this so that we can continue to collaborate together.

Tucker Max, Meghan McCracken, and the whole Scribe team made this dream project become a reality. I thank you all for all the frictionless fun times that brought it to fruition.

Ray Braun's magnificent illustrations brought the humanity of these somewhat otherworldly intellects into sharp focus.

Stuart Volkow is the brain behind the *Into the Impossible* podcast at UC San Diego where many of these interviews took place. His expertise, curiosity, and attention to detail turned each video and audio interview into a true treasure!

Jay Wujun Yow helped salvage some of the audio when all hope seemed lost.

Thanks to Tim Ferriss for his *Tribe of Mentors* project, which was a big inspiration behind this book.

Thanks to Eric Weinstein for assistance preparing questions and research with several of my guests on the podcast and showing me that sometimes it is the individual laureate who bestows honor upon the Prize, not the other way around.

My wife Sarah helped me in innumerable ways, as she always does. I cannot think of a way to repay her, but that won't stop me from trying. My kids inspire me to ask better questions and to never answer a question with, "Because I said so!"

Finally, if there was a golden award for gurus in the self-help genre, it would go to James Altucher, the man who inspired me to "choose myself" and write this very book, convincing me it was a repository of wisdom that was needed now more than ever. James is a scientist and a seeker. Thank you for choosing me to spend some of your time with on this Pale Blue Dot! Someday we will discover that elusive theory of everything that will allow humanity to *Skip the Line* and avoid total devastation. Until then, I will revel in your insight and wisdom and generosity of spirit and use it to try to become a better human being.

ABOUT THE AUTHOR

BRIAN KEATING is the Chancellor's Distinguished Professor at the University of California San Diego and the author of more than 200 scientific publications, two US Patents, and the bestselling memoir *Losing the Nobel Prize*. Keating did research at Case Western Reserve University, Brown University, Stanford, and Caltech. In 2007 he received the Presidential Early Career Award for Scientists and Engineers from President Bush. Keating is a Fellow of the American Physical Society, and co-leads the Simons Array and Simons Observatory cosmology projects in Chile. He is a pilot and an honorary lifetime member of the National Society of Black Physicists.

Printed in Great Britain
by Amazon